Luciana Maria Da Cruz Clavijo

The Urban Space

Luciana Maria Da Cruz Clavijo

The Urban Space

Relations between space and perception of violence

ScienciaScripts

Imprint
Any brand names and product names mentioned in this book are subject to trademark, brand or patent protection and are trademarks or registered trademarks of their respective holders. The use of brand names, product names, common names, trade names, product descriptions etc. even without a particular marking in this work is in no way to be construed to mean that such names may be regarded as unrestricted in respect of trademark and brand protection legislation and could thus be used by anyone.

Cover image: www.ingimage.com

This book is a translation from the original published under ISBN 978-620-6-76211-9.

Publisher:
Sciencia Scripts
is a trademark of
Dodo Books Indian Ocean Ltd. and OmniScriptum S.R.L publishing group

120 High Road, East Finchley, London, N2 9ED, United Kingdom
Str. Armeneasca 28/1, office 1, Chisinau MD-2012, Republic of Moldova, Europe
Printed at: see last page
ISBN: 978-620-8-28616-3

LUCIANA MARIA DA CRUZ CLAVIJO

THE URBAN SPACE:

RELATIONSHIPS BETWEEN SPACE AND THE PERCEPTION OF VIOLENCE

2024

SUMMARY

Breaking with the traditional approach to criminal studies, which was limited to the individual and legal aspects, the spatial dimension of the phenomenon of violence and crime has revealed the role of space in understanding and preventing the problem. Although it is not determinant, this dimension is crucial and its analysis must go beyond the simple spatial distribution of criminal occurrences. This study discusses the relationship between crime, violence and geographical space. To this end, the perception of geographic space was used as a relevant source for understanding the problem, as it takes into account the participation of people who deal with the problem on a daily basis. Based on the assumptions of currents in criminological studies that emphasize the role of space in crime studies, such as Environmental Criminology, it is believed that spatial characteristics, in addition to intervening in the criminal act, also interfere in people's sense of security.

Keywords: Space, Crime, Perception, Violence.

SUMMARY

1 INTRODUCTION

Violence and crime are historical problems, present in different forms of society. There are many theories that seek to explain criminal behavior from different perspectives. Some put the concentration of economic problems as the cause, others associate the difficulty of forming social ties in the community and within the family as predictive of criminal behavior. There are also those that focus on issues related to the inefficiency and selectivity of the justice system, among others.

Traditionally, criminological studies have focused on understanding criminal behavior from the perspective of the individual and their motivations, but some scholars have decided to broaden the explanatory focus of crime beyond the victim and the offender, also looking at the context in which the crime occurs. It wasn't until the beginning of the 20th century that space came to be considered an important dimension for understanding and preventing criminal phenomena.

There is an endless number of studies on crime and violence in various fields of knowledge, and spatial studies are emerging as a promising branch in this area. As such, geography researchers need to assume their role as spatial thinkers and contribute more actively and creatively to the search for understanding and solutions to the problem. Few geographic studies take an approach that goes beyond the spatial distribution of occurrences and analyses based solely on the use of geotechnologies supported by secondary data, thus limiting the potential for spatial research that can elucidate questions such as:

What factors could be associated with the unequal distribution of crime in space?

Does geographical space interfere with the territorial management of security?

How do people (residents and non-residents) perceive safety in the area and in the city?

In addition to incidents, what factors influence the perception of safety in the area?

Does the spatial dimension interfere with this perception?

Reflecting on these issues, this work aims to understand whether (and how) geographical space interferes with the occurrence of crime, the territorial management of security and people's perception of security. We believe in the active nature of space, which at the same time as it is used and transformed, acts as a conditioning factor for human actions. The definition of space proposed by Santos (2008), as an inseparable set of systems of objects and actions, will guide this work as it manages to encompass the complexity of human-environment relations by considering both spatial materiality and immateriality. The specific objectives include:

Investigate whether urban infrastructure is related to crime occurrences;

Evaluate the territorial management of public security services and its difficulties;

To find out how the space is perceived by people in terms of safety;

To detect whether the characteristics of public spaces interfere with people's sense of security.

The hypothesis tested here is that spatial characteristics related to urban infrastructure interfere not only in the occurrence of crime, but also in people's spatial perception of safety, which leads us to believe that fear of crime can often be associated with space and not with the occurrences of crime in space. As a result, some spaces end up being "criminalized" or "stigmatized" in the process of constructing mental maps of the city.

The issue of violence and crime in Brazil is one of the biggest obstacles to the well-being of the population, which requires huge financial investments in the area of public security.

In the city of Recife, which has already topped the ranking of the most violent cities in the country, security is one of the main demands of the population, who are frightened by violent crime. However, in order to find solutions to this problem, it is necessary to understand the causes, effects and conditions that lead to criminal and violent behavior. To do this, we need a base of information and research that can elucidate issues and support prevention policies in the search for solutions to the problem. However, official data is insufficient to draw up a profile of urban violence in the city, since this data is only a snapshot of crime reports and not of the reality experienced in the spaces where they occur.

This makes it necessary to complement the data with means that take into account not only the numbers, but also the opinions and perceptions of those who deal with the various forms of violence and crime on a daily basis.

It is therefore believed that the problem in question should not only be thought of from the top down, but also from the bottom up. And the aim of this work is to serve as a source of empirical evidence on the spatial conditions (physical and human aspects) of the city in general and to investigate the space-crime relationship in the field.

A methodological proposal is the triangulation of methods, which uses both quantitative and qualitative approaches, enabling a more consistent analysis of the problem. Although this methodological proposal does not guarantee a complete understanding of the phenomenon and poses some challenges, it does provide a more comprehensive view of the criminal phenomenon.

This work is divided into four chapters. The first seeks to define key concepts such as violence and crime, moving on to arguments about the role of the media and the fear of violence, and the association between crime and poverty, ending with a brief overview of the issue in the country.

In the second chapter, an approach was made to geography in criminal studies, explaining the contribution of currents such as the Ecology of Crime and Environmental Criminology, moving on to an explanation of the methods and techniques adopted in spatial studies of crime and, finally, an attempt was made to demonstrate how the categories of geographical analysis relate to the problem and how the perception of space can serve as a tool for this type of study.

The third chapter reveals the research methodology, whose procedures began with an exploratory analysis of data from the 2010 Demographic Census, resulting in the creation of an index on the characteristics of the surroundings of households (ICED). Next, the existence of correlations between the index and crime data was ascertained, the results of which led to qualitative research techniques, through field observations and interviews, aimed at capturing people's spatial perception of security.

The fourth and final chapter presents the results at different scales, from municipal to intra-neighborhood. The aim of this methodology was to demonstrate that the issue of safety has a dimension that goes beyond the location and spatial distribution of incidents, and must also take into account the physical (infrastructure) and human (perceptions) aspects present in these spaces.

2 VIOLENCE AND CRIME: SOME NOTES

When talking about the "state of the art" of research, the feeling that seems to invade researchers is that they don't know the totality of the studies in a given area of knowledge (FERREIRA, 2002), especially when dealing with a subject as complex and debated as violence and crime. Aware of the researcher's limitations in this process, a selection of works in various areas of knowledge was carried out in order to glimpse the main issues and approaches on the subject that make it possible to understand the relationship between space, crime and the perception of violence.

In this chapter, we sought to clarify some concepts and highlight some contributions from scholars in the field, debating the association between crime and poverty, which leads to the criminalization of areas of poverty, and the role of the media and the fear of violence in this problem. Finally, there is a brief overview of the issue in the country.

2.1 VIOLENCE AND CRIME: DIFFERENCES AND DEFINITIONS

Despite the usual indistinct use of the terms, violence and crime have different meanings. Not every form of manifestation of violence is considered a crime and not every crime is violent. For example, theft is a type of crime that is not considered violent and authoritarianism can be considered a form of violence, but it does not constitute a legally prescribed crime.

One definition of violence that deserves mention is that suggested by Arendt (1994), for whom violence represents the absence of power, a resource or control tool for exercising unrecognized or legitimized power. Arendt's approach is interesting for understanding that violence is rather a political problem, which leads us to a form of violence that is intangible and difficult to identify, but of vital importance, because as Santos (2000) makes clear, structural violence is the basis for the production of other manifestations of violence. From this perspective, Pedrazzine (2006) proposes the following reflection on violence in the urban space:

> We assume that the hegemony of the metropolis, the process of ultra-urbanization, cannot take its course without exercising violence against the territory - against its "nature", its topography, its character - and against its inhabitants. Would it therefore be correct to think of the violence of just a few inhabitants (for the vast majority) as a response to the violence of urbanization, urban society, fragmented territory, the economy, inequality and segregation? (PEDRAZZINE, 2006, p.54)

In this case, violence would be the result of a dialectical process between structural violence and its derivative forms (crimes). While acknowledging the relevance of this reflection, Zaluar (1999) warns that the main difficulty with this type of approach is that violence becomes synonymous with inequality, exploitation, domination, exclusion, segregation, which distances us from actions characterized by excessive or uncontrolled use of physical force (or its instruments) in social interactions. And so, this author proposes looking for a definition of violence in the etymology of the term, which comes from the Latin "violentia" and refers to "vis" (strength, vigor, the use of physical force or the resources of the body to exert its vital force), and clarifies that this force "becomes violence when it exceeds a limit or disturbs tacit agreements and rules that order relationships, acquiring a negative or harmful charge" (ZALUAR, 1999, p.8).

The concept of crime, on the other hand, is less comprehensive and is restricted to actions codified in criminal legislation. For sociologist Émile Durkheim, crime has a character of normality because it is present in all societies and is conceived as the transgression of socially imposed rules (RATTON, 2009). From this perspective, it is the collective values that define the rules, determining right and wrong, which is why there is talk of the legality and legitimacy of crime, i.e. crime is legitimate when socially recognized as such and legal when provided for by law. The convergence of these factors varies according to society and historical context.

In Brazil, according to DECREE-LAW 3914/41 in its first article, a crime is considered to be "a criminal offense for which the law imposes a penalty of imprisonment or detention, either alone or alternatively or cumulatively with a fine". Therefore, when working with official criminal records data, we are dealing with crime, which can present varying degrees of violence as a resource. On the other hand, when dealing with people's perceptions of crime, the terms are confused, as the degree of violence employed may be more relevant to people than the legal transgression.

2.2 POVERTY AND VIOLENCE AND THE CRIMINALIZATION OF SPACE

Even though this issue has been exhaustively debated and discredited in academic circles, the association between crime and poverty seems to persist in common sense. Although the hypothesis linking poverty and violence has no directly identifiable advocate (MISSE, 2006), it continues to be part of the debate.

According to this hypothesis, violence is a consequence of accelerated urbanization, which has led to a series of social problems such as unemployment, slums, etc., leaving a contingent of the population on the fringes of the criminal world. From this point of view, poverty and the consequent exposure to social risk are presented as the main causes of routine crime in urban centers, especially crimes of greater offensive potential and crimes against property, which, in turn, make up the majority of recorded crimes (OLIVEIRA, 2008).

When considering the complexity and scope of the terms, as discussed in the previous topic, it is known that violence and crime cannot be reduced to a single cause, as these phenomena are linked to various factors. Even the definition of what can be considered "poverty" is controversial.

With the clarification that the obstacles to human development lie in the unequal opportunities for choice, rather than in poverty itself, a new concept emerges for understanding criminal behavior, denying a direct relationship with poverty. Zaluar (2007) assures us that the discourse associating violence and poverty is insufficient to understand, for example, the armed conflicts that kill so many young men in the country, and that there are new global forms of illegal and violent economic activities (drug trafficking) that cannot be considered simple survival strategies for young people who die before the age of 25.

Furthermore, this bias reinforces discrimination against the poor, both in the institutions in charge of repressing criminal behavior and in the imagination of the general population. For this reason, the author recommends that when dealing with the relationship between poverty, crime and violence:

> From the perspective of complexity, we need to discuss how poverty and the lack of employment for poor young people relate to the mechanisms and institutional flows of the justice system in its ineffectiveness in combating organized crime. This crosses all social classes and is connected to legal businesses and governments. (ZALUAR, 2007, p.35)

In this context involving the lack of control over crime and violence in the country, society lives in a constant dilemma between freedom and security. In his work "*A oficina do Diabo e outros estudos sobre criminalidade*" Coelho (2005, p. 286) explains that "the marginalization of crime consists of attributing to a certain class of behaviour high probabilities that they will be carried out by the type of individual who is socially marginal or marginalized".

Pedrazzine (2006) mentions the issue by saying that, while it is said that the city's poor are violent, the attention given to the violence they suffer is inverted. The violence of living near luxury condominiums and fortified mansions, often without access to basic goods to guarantee reasonable living conditions, is forgotten.

Sá (2010, p. 120) argues that "despite the repulsion of many individuals towards the violent actions of poor communities, we cannot hide the fact that, in a way, they express resistance to socio-spatial dissymmetries".

However, crimes committed by the poor tend to have a greater impact on the population than crimes committed by the better-off. Misse (2006) explains:

> "Poor people's crimes" are those characterized by the presence of violent means in their execution, such as robbery, extortion by kidnapping, bodily injury, homicide, etc., while "rich people's crimes" are those characterized by the presence of deception, such as fraud and corruption. The fundamental difference in these crimes is that those "said to be of the poor" cause greater social repudiation, not because they are related to poverty, but because they cause greater physical and moral damage to the victim (MISSE, 2006, p. 22).

The question then arises: why are violent crimes more common among the poorer population? This is certainly not an easy answer and would require an analysis that is not part of the objective of this work. However, some efforts in this direction are worth mentioning, such as the theory proposed by Machado da Silva (2008) on "violent sociability", which refers to a context in which the form of social organization is permeated by violence.

We must also not forget a very important factor that differentiates individuals in their criminal practices: power. After all, as noted by Melo (2010):

> What are the chances that a senior executive, in order to obtain easy and illicit money, will use a gun and rob people at traffic lights? [...] what are the chances that an unemployed person who hasn't completed basic education will have access to an account abroad and use it to embezzle funds from a public tender, social security or taxes? (MELO, 2010, p. 114).

In other words, even in the world of crime, social inequality becomes a condition. It's clear that policies to combat poverty are essential to guarantee decent living conditions for the population, but this doesn't mean that this will lead to a reduction in crime. Back in 1988, Paixão highlighted the Brazilian paradox that the increase in social indicators was accompanied by an increase in crime rates in large urban centers. In this regard, Beato and Reis (2000, p.10) clarify that "social and economic development, contrary to what is imagined, can be a more favorable context for the growth of crime rates, especially in the form of crimes against property".

There is a large following of those who believe that a large part of crimes can be explained by economic motivations, but the examples almost always fall on crimes committed by people living in poverty, which in these cases are considered illegitimate mechanisms for the consumption of goods, the desire for which is socially stimulated, as proposed by Robert Merton's Social Anomie Theory (1938). So what about white-collar crime? Aren't these the biggest economically motivated crimes? Soares (2000) explains that "those who attribute involvement in crime to economic needs often forget the role that culture, values, social norms and symbols play. Self-esteem is as important to human survival as a plate of food" (*apud* MACEDO et. al., 2001, p.517).

One problem that helps to explain the persistence of the association between crime and poverty is related to statistical data on crime. According to Carreira (2009) presenting data from the Ministry of Justice, in 2006 Brazil had a prison population that was 95% poor or very poor. Coelho (2005) points out a series of flaws in this data and explains that the fact that the vast majority of crimes recorded (registered, judged, etc.) involve people on low incomes can be explained by a possible inefficiency in registering and punishing white-collar crimes.

Thus, due to the coinciding relationship between indices indicating poverty (income, schooling, etc.) and crime statistics, the belief that criminals come from the socially marginalized population is reinforced.

The selectivity of the justice system, which directs crime repression actions towards poor areas and their population, would then be a counter-argument to this type of association, which leads Adorno (1994, p. 149) to state that "if crime is not a class privilege, punishment seems to be". However, Oliveira (2008) argues that due to the density of poverty in the country, it would be abnormal to have a majority of rich people in prison, in other words, he argues that, in proportional terms, the data shows a certain normality.

In addition to these questions about the reliability of crime data, in Brazil there is a difficulty in obtaining such data for research purposes, often justified by the need for secrecy of the information and, when it is released, it is not very clear to the majority of the population. So the question is: can the disclosure of crime data hinder security services? In some US cities, Chicago for example, the release of data has become a tool in the fight against crime (this will be exemplified in the next chapter in the subtopic on mapping).

Some efforts have been made to standardize information on crime in the country and create a database that would allow for more reliable comparative studies (LIMA, 2011a). However, it can be seen that

> [...] in parallel with the necessary investment in structuring tools and technologies to reverse the scenario of worsening information quality, Brazil faces the challenge of agreeing on transparent and mandatory rules for recording and publishing data in the area. After all, the introduction of information management technologies alone does not produce organizational or substantive changes in the process of managing and using public security data. (LIMA, 2011a, p.9-10)

In any case, it must be understood that crime statistics represent summaries built on observations, which are more a portrait of the social process of reporting crimes than an accurate picture of the universe of crimes actually committed in a given location (SAURET, 2012).

The criminalization of the poor, already discussed by anthropologist Janice Perlman in 1977, leads to the criminalization of areas of poverty, i.e.

favelas and neighbourhoods lacking in infrastructure that house the low-income population. These are areas that not only suffer from other problems (lack of sanitation and garbage collection that cause the proliferation of diseases, inadequate housing and lighting and in many places in Brazil correspond to areas at risk of landslides and floods because they are the result of irregular occupation, etc.) but also suffer from a lack of security and, in some cases, are dominated by drug trafficking. These are areas that usually concentrate crimes against the person (threats, bodily injury and homicide), a common pattern in Brazilian cities.

Following Edmundo Campos Coelho's (1980, p. 291) assertion that the association between crime and poverty is "methodologically fragile, politically reactionary and sociologically perverse", I would add that it is also geographically "deterministic" in imputing a potentially criminogenic property to areas of poverty.

Poor neighborhoods and shantytowns are stigmatized as dangerous environments, and this perception is due not only to prejudice against their residents, but also to the poor conditions of local infrastructure and sanitation, since these environmental conditions, which indicate poverty, are perceived as places forgotten by the government, where a culture of violence and revolt prevails.

This discrimination against spaces that are home to the poor is reinforced by the way the media explores the issue of violence. The emphasis given to the names of some poor neighborhoods and communities in the media contributes to the construction of mental maps about the city. People who have never been to a certain neighborhood may consider it dangerous because of what they see on the news, thus bringing out the prejudice that associates crime with poverty. As a result, warns Luna (2013), poor neighborhoods and favelas are physically and symbolically excluded from the rest of the city.

As explained above, there are various arguments that contest the association between crime and poverty, however, the statistical data and crime repression measures corroborate this idea, which leads Peralva (2000, p. 81) to argue that it is difficult to discredit this relationship when "the geography of violent deaths is concentrated in poor peripheries and not in rich neighborhoods; the geography of police interventions, or the population of prisons, suggest that the association between crime and poverty is indisputable". However, due to questions about the reliability of criminal data, which will be discussed below, and the existence of a range of crimes that are not officially recorded, this association can be contested and other means of getting to know these areas should be sought.

2.3 THE MEDIA AND THE FEAR OF VIOLENCE

According to Bauman (2008), "Fear" is the name we give to our reaction to a threat, caused by "uncertainty" or "ignorance" of how to act to face it. It is a feeling common to humans and animals, but the author clarifies that humans know a type of social fear, called "second-degree fear" or "derived fear", which guides behavior, whether or not there is an immediate threat. According to the author, this secondary fear:

> It can be seen as a trace of a past experience of facing the direct threat - a remnant that survives the encounter and becomes an important factor in shaping human conduct even if there is no longer a direct threat to life or integrity (BAUMAN, 2008, p.9).

Although many of today's fears already existed in remote times, today the fear of violence and crime combined with other fears have established a new form of social interaction that has been changing habits and fueling feelings of insecurity in the country. Adorno and Lamin (2006) state that:

> The fear of crime, expressed through narratives and speeches, opinion polls and victimization surveys, concerns very deep collective feelings, rooted in the most recondite realms of consciousness and the collective imagination, overlaid by layers bequeathed from generation to

generation by historical time (ADORNO and LAMIN, 2006, p. 169).

Souza (2008) adopted the term "Phobopolis" to refer to cities dominated by the fear of violent crime, in which fear and the perception of growing risk assume an increasingly prominent position in conversations, in the mainstream news, etc., the latter being the main means by which the population has access to information about violence and crime.

On a daily basis, TV news, radio, the internet, newspapers and magazines broadcast the various cases of murder, robbery, kidnapping and aggression that occur around the world. There are also specialized programs on the subject that explore the cases on different scales. In this way, the media's role in shaping the public's imagination about the issue is undeniable .[1]

In her book "*Histórias que a mídia conta: o discurso sobre o crime violento e o trauma cultural do medo*". Patrícia Bandeira de Melo (2010) seeks to understand how the process of fabricating meanings about violent human action, which occurs in the discourse produced by the Brazilian press, contributes to establishing a sense of fear and determining behavior among individuals. She states that:

> The fear of violent crime in Brazil stems from a projection that individuals make of real problems. However, the design of the episodes, their extent and the way they are conveyed in journalistic narratives makes the feeling of insecurity grow even more (MELO, 2010, p.146).

However, she clarifies that there are no conclusive studies that allow us to affirm a significant correlation between the dissemination of news about violent crimes and the actions of individuals, and also that it cannot be said that the fear arising from journalistic discourse is an intentional meaning effect of the media. However, he assures that:

---

[1] - Regarding the notion of the imaginary, Durand explains that it is "[...] a set of images and relations of images that constitutes the thought capital of Homo Sapiens" (1997, p. 18), including all types of human representation through signs, perception, etc.

> Media narratives, by constructing meanings about violent human action, the perpetrators of evil and the imminent risk of any individual becoming a victim of a violent crime, elaborate meanings for individuals that suggest the perception of generalized fear as a meaning effect, as collective trauma" (MELO, 2010, p 21).

Of course, individuals are not just recipients of news; on the other hand, they also influence the media by seeking out this type of information, as the media seeks to expose what arouses the public's interest.

It is also worth pointing out that, although it is true that crime is a real element of the Brazilian social structure, what the press highlights is not routine crime, but the unusual, the improbable. In other words, there are violent crimes that, even if rare, get space in the media, while others that have a high statistical incidence remain invisible and, as a result, the official rate of violent crime is not associated with the volume of news on the subject (MELO, 2010).

In this way, people's mental maps of the city and its risk areas are constructed with the influence of the media and not statistics. What's more, many of the crimes exploited by the media today are caused by tensions in intersubjective relationships that seem to have nothing in common with everyday crime, as Adorno (2002) makes clear:

> These are an infinite number of situations, generally involving conflicts between acquaintances, the outcome of which often ends accidentally and unexpectedly in the death of one of the contenders. They include conflicts between partners and their partners, between relatives, between neighbors, between friends, between work colleagues, between acquaintances who frequent the same leisure areas, between people who cross paths every day on public roads, between bosses and employees, between shopkeepers and their customers. [...] More often than not, they reveal how sensitive the social fabric is to tensions and confrontations that, in the past, didn't seem to converge so abruptly towards a fatal outcome (ADORNO, 2002, p.100).

However, it is the effect of this information that fosters fear in the population, even in individuals who have not been directly victimized. Crime, especially violent crime, has an impact on various aspects of social relations, such as: the logic of urban planning (residential segregation, fortified housing, preference for private environments, etc.); the behaviour of the population (fear that leads to changes in habits); social relations (new forms of socialization, compromised citizenship); the tightening of security and surveillance, all of which culminates in a reduction in the population's quality of life (CARRIÓN, 2008).

2.4 VIOLENCE, CRIME AND PUBLIC SAFETY BRAZIL

The issue of violence and crime in Brazil is one of the biggest obstacles to the well-being of the population. In the global context, it is one of the countries with the highest number of firearm-related deaths, ahead of countries with much larger populations such as China and India (WAISELFITSZ, 2013).

Security appears among the main concerns of Brazilians in different opinion polls (IBOPE, 2010; DATASENADO, 2010; INSTITUTO AKATU, 2009). In 2012, 56,336 homicides were recorded in the country, corresponding to a rate of 29 homicides per 100,000 inhabitants.

In analyzing this evolution of data on lethal crimes in the country, Waiselfisz (2014) shows some changes in the spatial dynamics of the rates. His research revealed that states with a large demographic weight (São Paulo and Rio de Janeiro) and high homicide rates (Pernambuco, Espírito Santo, Mato Grosso, Mato Grosso do Sul and the Federal District) showed a drop in rates, pulling the national rates down, especially São Paulo and Rio de Janeiro.

On the other hand, it also reveals that 20 of the country's 27 states significantly increased their indicators. Alagoas and Pará had a sudden and strong eruption of violence, more than doubling their rates in the period, and came to occupy a prominent place in the national context at the end of the decade (1st and 3rd place in 2010, respectively). The states of Ceará,

Goiás, Bahia, Sergipe and Paraíba also stood out in the decade analyzed. Thus, at a regional level, the highest rates in recent years have been observed in the Northeast.

The dynamic centers of violence have thus shifted from a small number of large cities to many medium-sized or small municipalities. Other characteristics presented by the research are that this type of violence predominantly affects young, black and poor people (WAISELFISZ, 2014).

Data from IPEA (2009) show that the feeling of insecurity among the population in the country was 78.6% among adult Brazilians who are very afraid of being murdered and 73.7% who are very afraid of being victims of armed robbery (SIPS, 2010)[2] .

According to a sample victimization survey carried out in 2009 by the IBGE, the declarations made it possible to state that, as the population moved away from home, the feeling of security decreased.

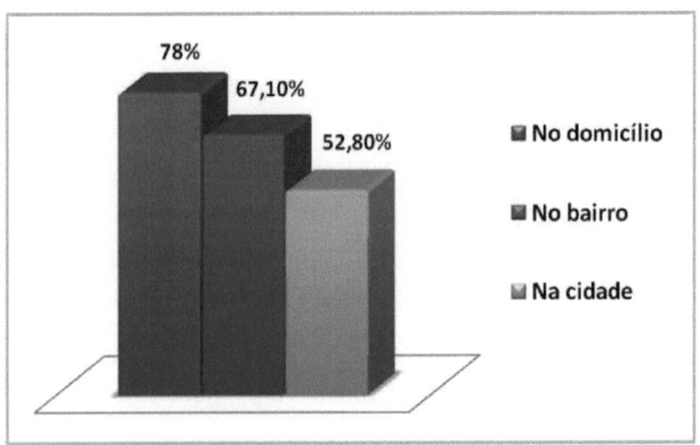

Graph 1: Brazilians' sense of security by scale of comprehensiveness
Source: IBGE - PNAD 2009 / Graphic editing: the author

The victimization survey also revealed that the feeling of safety varies according to age, being more present among children and

---

[2] - Presidential Message PPA 2012-2015.

adolescents. In terms of gender, men declared themselves safer than women, and when it comes to income, there is a higher proportion of safety in the population with a higher monthly household income *per capita*. Analysis of *per capita* monthly household income classes in relation to those victimized by physical aggression also revealed that the percentage of people attacked grew in the opposite direction to income.

On the other hand, when analyzing the crimes of robbery and theft, it was observed that the higher the income class, the higher the percentage of people who were victims of the crimes in question. Of the 58.6 million permanent private households, around 60% used some kind of security device[3] . In all regions, this percentage exceeded 50%, especially in the Southeast with 63.9% and the Midwest with 64.9% (PNAD, 2009).

Among the researchers who pioneered the issue of crime and violence in Brazil, some stand out: Alba Zaluar, Antônio Luís Paixão, Claudio Beato, Edmundo Campos Coelho, Michel Misse, among others, who helped to form a field of theoretical and methodological guidelines that influence research on the subject to this day (LIMA and RATTON, 2011b). These authors brought to light the fragilities of the association between poverty and crime, as well as other peculiarities surrounding the issue in the country.

Studying urban violence, Zaluar (2007) observed that it grew sharply at the time when the country was beginning its process of democratization. In this way, she considers that the boom in violence in Brazil was accompanied by the diversification and modernization of the economy. However, the author clarifies that there was no democratization of political and legal institutions to accompany this modernization and thus considers that these spheres, specifically the functioning of the justice

[3] - The devices considered were: window/door grill; peephole, door opening, chain on the door latch or intercom; electrified fence, wall or railing over 2 meters high or with shards of glass or barbed wire, and/or electronic alarm; extra locks and/or bars on the door/window against break-ins; dog; video camera; private security and/or gate and other security mechanism.

system, constitute the "hard core" of discrimination in Brazil, which violates the rights of the poorest.

Coelho (2005) asserts that the roots of violence in Brazil stem from its socio-economic and territorial formation characterized by the concentration of land and income, and that the formation of a reserve army that was not absorbed into the labour market led to the emergence of the so-called "dangerous classes"[4] , because given the country's marked socio-economic inequality, it is common to think of crime as an alternative for the dispossessed, a way of attaining the goods and services that everyone is encouraged to obtain. In addition, there are some theories that seek to explain what is peculiar about Brazilian society that, in a way, underlies deviant behavior. Starting with the "habit" that most Brazilians have of circumventing the law for their own benefit. Melo (2010) explains this:

> [...] throughout Brazil's history, illegal practices have been valorized through the *Brazilian jeitinho* - that is, the tendency to take advantage and try to act above the rules. The link between crime and culture is made through a connection between criminal practices and a perception of national identity as the result of a society in which roguery is a mark of the Brazilian, thus leading to the construction of a social structure whose matrix is crime as an element of the country's cultural tradition. (MELO, 2010, 160).

Another issue also linked to culture is highlighted by Maria de Carvalho Franco (1997) in her work *"Homens livres na ordem escravocrata" (Free men in the slave order)*. When studying the population of the interior of São Paulo, the author draws attention to cultural aspects that have a strong connection with violent behavior:

> The actual behavior of the people involved in these disputes corresponds exactly to the requirements of bravery that they proclaim. [...] When personal attributes are called into

---

[4] - This was how impoverished urban workers were considered, especially at the beginning of the Republic, when they were subject to social control that included illegal arrests, mistreatment in police stations and arbitrary persecution, among other things (ADORNO, 2002).

> question, there is no other socially accepted recourse but to strike back in order to restore the integrity of the aggrieved party. [...] Violence thus becomes legitimate conduct (FRANCO, 1997, p. 51).

This relationship is also seen in the interior areas of the Northeast, where many crimes are "justified" by issues related to honor. Waiselfitsz (2013) speaks of a "culture of violence" as one of the factors responsible for firearm-related deaths in the country, noting the high proportion of murders resulting from frivolous motives such as fights, jealousy, conflicts between neighbors, disagreements, arguments, domestic violence, traffic disagreements, etc.

Zaluar (1999) highlights the issue of symbolic associations related to the use of firearms, money in the pocket, the conquest of women, etc., as a way of linking violence to an "ethos of hypermasculinity" that seeks recognition through the imposition of fear.

However, Beato (2012) warns that culturalist theories often rest on the false assumption of the 'complete socialization' of criminals and young people involved in illicit activities, and can lead to the idea that people living in the same area uncritically share local values.

The subject of Public Security, much debated in the country, can be understood as "the narrower aspects of crime and violence as defined by law, and which are the object of action by the organizations that make up our criminal justice system" (BEATO, 2102, p 26), in other words, it is limited to the crimes provided for by law and the policies to combat and prevent them, rather than a broad discussion of violence. The aforementioned author also warns that public security policies in the country are based on two convictions: the first stems from the belief that crime is the result of socio-economic factors that block access to legitimate means of earning a living, from which arise proposals for social reforms and the re-education and re-socialization of individuals to live in society. And the second is the belief that crime finds ideal conditions to flourish when individual discipline and respect for norms are low, resulting in policies of more repressive measures of social control by the state apparatus (BEATO, 2012).

According to Sapore (2007), public security policies in the country have been marked by a common characteristic: crisis management due to the lack of a more systematic managerial rationale. As a result, the problem was dealt with in a superficial way, with only a repressive and immediate nature, until the first National Security Plan was drawn up in 2001, based on the human rights discourse, which certainly changed the perspective of working on the issue in Brazil. Although the discourse did not come to fruition in practice, this plan represented a step forward both in terms of incorporating the issue of security more effectively into the federal government's framework and in terms of emphasizing the importance of preventing violence.

In 2007, the National Program for Public Security with Citizenship (PRONASCI) was launched. This program announces a new proposal for territorial management that goes down to the municipal level, valuing their contribution, which until then had only been the responsibility of state governments. Some of the efforts of new management models that assign responsibilities to municipalities in the area of security are aimed at adapting policies to local realities. However, the debate about the inclusion of the municipality as a territorial scale for security management and the effectiveness of these practices remains "timid" in the country.

From a spatial perspective, the programme has made some progress, with its guidelines highlighting guaranteed access to justice, especially in vulnerable territories, and measures to urbanize and recover public spaces, highlighting the territory as one of its main focuses and proposing geographically targeted actions. However, the systematization of information and dissemination of actions at the federal level remains weak, with state initiatives remaining isolated.

Brazil is also marked by diversity when it comes to crime, and although the manifestations of the metropolitan regions are taken as paradigms for the country as a whole due to the disorganized process of urbanization in the big centers, there are also specific characteristics within the big cities and, as seen above, significant regional variations.

Beato (2012) also warns that when dealing with the issue of security, it is necessary to consider some aspects of crime, such as the rationality of choosing the world of crime in the face of high impunity and the difficulty of social ascension in the country, in which it may seem advantageous to weigh up the costs and benefits of crime in relation to legal means. For this reason, he argues that "increasing the benefit of not committing crime should also be considered through measures to encourage the use of legitimate means of earning a living" (BEATO, 2012, p 43). He also warns that many of the rewards sought by criminals are not expressed in material terms, and that the complex of punishments is complicated and cannot be easily generalized.

In short, there are a number of causes and intentions that lead to deviant behavior and it is also necessary to consider some of the mechanisms behind many criminal actions, such as political deals, power games and corruption schemes that need to be combated as much as the more obvious factors. And consider that, as Beato (2012) states, some factors of crime are not under the control of the state and may be linked to other forms of intervention.

3 GEOGRAPHY IN STUDIES ON CRIME AND VIOLENCE

Since the Second World War, geography has multiplied its focus of interest, drawing closer to other human sciences and revealing space as a social production, thus becoming the science that studies the process of spatial organization of society (LAVILLE and DIONNE, 1999). However, geographers' contributions to studies on crime and violence only began in the 1970s . [5]

According to Harries (1999), spatially-oriented studies of crime have been carried out by sociologists and criminologists since the 1830s. The first criminological studies whose spatial dimension was highlighted were attributed to the French (GUERRY, 1833; QUETELET, 1842), however, this dimension became more visible through the studies carried out by researchers from the Chicago School.

In order to understand how geographical space fits into the problem of crime and violence, we need to evaluate the contributions that have already been made in this area. For this reason, we will begin this chapter with a survey of spatial approaches in criminal studies through the currents of Crime Ecology and Environmental Criminology, which pay attention to aspects relating to the economic, social and physical structure of cities. This is followed by a discussion of some of the methods and techniques applied in geographical studies of crime, highlighting spatial analysis and mapping. The chapter ends with considerations on the categories of geographical analysis, the spatial perception of violence and its use as a methodological tool.

---

[5] Among the studies carried out during this period, Harries (1999) highlights: Harries (1971, 1973, 1974), Phillips (1972), Pyle *et al.* (1974), Lee and Egan (1972), Rengert (1975), Capone and Nichols (1976), among others.

3.1 ECOLOGY OF CRIME

The Ecology of Crime refers to the study of areas with a high incidence of crime, looking for relationships between the individual and the surrounding environment. This type of study gained visibility through researchers from the Chicago School. The Social Disorganization Theory (SDT) is the most emblematic ecological approach to crime. Although it was created in the mid-1920s, it influences work of this nature to this day. This theory was developed by Clifford Shaw and Henry McKay (1942) during their studies of juvenile delinquency in the city of Chicago.

These researchers showed by comparing different time series that the distribution of delinquency rates had a strong correlation with certain socio-demographic factors (economic conditions, composition and turnover of the population, etc.) and that they were highly concentrated in certain areas of the city. This research served to refute the association between delinquency and race, since the areas where these rates were concentrated showed ethnic heterogeneity. This socio-demographic heterogeneity in some communities was then seen as responsible for creating an environment conducive to the development of delinquency.

Therefore, one of the assumptions defended by this current is that delinquency is not only caused at the individual level, but is also a response of normal individuals to abnormal environmental conditions (WONG, 2001). This association has acquired a deterministic and fallacious character, since to see space as a defective environment that generates criminals is to neglect, for example, the propensity to crime of individuals from different social classes, as Sutherland (1940) pointed out when he drew attention to white-collar crime, or to ignore the individual's choice, as proposed by Cornish and Clarke's Rational Choice Theory (1986) which, in a way, can elucidate cases in which two individuals brought up in the same environmental conditions and where the same value system prevails opt for different behaviors.

Recent ecological models for studying crime seek to understand the multifaceted nature of violence and identify the factors that influence behavior, increasing the risk of committing or being a victim of violence. In this way, they try to answer why some localities have high crime rates (BEATO, 2012).

In an attempt to answer this question, Siva (2012) interpreted TDS as a theory that chooses the properties of community structures as determinants of the non-uniform distribution of crimes, characterizing some areas or neighbourhoods as violent. However, he warns that:

> Not all poor neighborhoods have high crime rates, but those in which ecological indicators of residential mobility, heterogeneity, family breakdown and chronic unemployment combine, there is a process that leads to the weakening or breakdown of formal and informal instances of control. In this case, the organizational capacity of residents is reduced and the likelihood of criminal behaviour is significantly increased (SILVA, 2012, p.42).

Most of the studies on SDT have been carried out in metropolitan areas (ROH and CHOO, 2008) since, due to the density and heterogeneity of the population and the various socio-spatial problems it brings with it, urban spaces have become the preferred field for criminal events over the years. And the concentration of crimes in these spaces has led some authors to speak of "typically" urban violence (PEDRAZZINE, 2006; CARRIÓN, 2008).

However, crime is not an exclusively urban phenomenon. With this in mind, American researchers Osgood and Chambenrs (2000) tested TDS in some rural areas of the United States. These authors defend the concept that SDT "is based on principles of community organization and social relations applicable to communities of all types and configurations" (OSGOOD and CHAMBERS, 2000, p.81), and thus sought to disassociate SDT from a specific geographical condition.

However, it is common in these studies to misconceive geographical space as a mere stage for human actions. Although space itself cannot be considered a preponderant factor, the way it is occupied, built and managed can facilitate or hinder human actions, including crime.

For this reason, this study considers that the association between the fear of violence and the lack of infrastructure in some places are also factors that lead to many spaces being considered dangerous. In this way, it can be said that, in the same way that human stereotypes are created about offenders, they are also created about the spaces within the city.

3.2 ENVIRONMENTAL CRIMINOLOGY

The term Environmental Criminology refers to the study of certain spatial aspects of crime, raising questions such as: where, when and how do crimes happen? (BRANTINGHAM's, 1981). To avoid the term being confused with Environmental Crimes (or Green Criminology), which is related to crimes committed against the natural environment (damage to water, soil, flora, fauna, etc.), the concept of environment used here corresponds to the "surrounding conditions" that influence the criminal phenomenon (including natural resources and built objects), as well as the specificities of a given locality (urban design, use and management of space).

Environmental criminology focuses on the motivating factors behind criminal events, emphasizing the relevance of the spatial component in understanding certain crimes. It assumes that certain environmental conditions can inhibit or facilitate criminal activity, and that criminal behavior is often dependent on the situational context.

According to Brantingham's (1981), the occurrence of a crime presupposes the inseparable existence of four elements (law, offender, victim and place), and they clarify that environmental criminologists:

> [...] they ask about the physical and social characteristics of the crime scene. They ask about the perceptual processes that lead to crime scene selection and the social processes of ecological labeling. Environmental criminologists also ask about the spatial distribution of targets and offenders in

> urban, suburban and rural areas. Finally, environmental criminologists question how the fourth dimension of crime interacts with the other three dimensions to produce criminal events (BRANTINGHAM and BRANTINGHAM, 1981, p.61).

Three important theories of criminology corroborate the environmental perspective, pointing to the importance of space in understanding crime: *Rational* Choice Theory; *Routine Activity* Theory and *Crime Patterns* Theory.

The first, Rational Choice by Cornish and Clarke (1986), suggests that most criminals weigh up the risks and benefits associated with the criminal act, selecting their targets and defining the means to achieve their goals. Although it examines crime from the point of view of the perpetrator, it provides grounds for highlighting the importance of place in this decision-making.

The second theory, Cohen and Felson's Routine Activity (1979), tries to explain the occurrence of crime through the confluence of some circumstances such as: the existence of a motivated aggressor, a desired target and the absence of capable guardians, in other words, it suggests a favorable circumstance in space and time for the occurrence of crime.

There are even studies that try to explain criminal behavior by relating routine activities to climatic phenomena. Cohn (1990) argues that:

> For example, during nice weather, people tend to spend more time outdoors, resulting in greater opportunities for personal interaction and increased availability of victims, as well as an increase in the number of empty (and therefore more vulnerable) dwellings. Inclement weather, such as cold ambient temperatures, reduces the number of people available as victims [...] However, those few individuals who are outdoors during unpleasant weather are more vulnerable, as there are fewer potential witnesses to stop the criminal (COHN, 1990, p. 52).

However, this is a relationship that is difficult to prove, has a deterministic charge and cannot be considered in places with less significant climatic variations.

Finally, the third theory, Criminal Patterns by Patricia and Paul Brantingham (1981), explores the interactions of criminals with the physical and social environment that influence their choice of targets, moving from simple description to modeling of the ways in which different socioeconomic situations and occupational and employment patterns affect the opportunity structures of crime in different spaces. Although they provide different explanations for the occurrence of crime, the first two focusing more on the individual and the latter on socio-spatial aspects, these theories are mutually supportive and lead to the belief that offenders do not choose their targets at random.

According to Eck and Weisburd (1995), just like any other individual, offenders move between activities such as school, work, shopping and leisure and, in the midst of these activities, they become aware of opportunities favorable to committing crime. In this way, the aforementioned authors state that "the offender will be made aware of only a subset of the possible targets available. [...] While some offenders may aggressively seek out unexplored areas, most will conduct their searches in familiar areas through non-criminal activities" (ECK and WEISBURD, 1995, p. 6), and argue that five points can help in understanding research into the importance of places:

- the concentration of crime on specific premises (e.g. bars);
- the high concentration of crime at some addresses and the absence of crime at others;
- the preventive effects of various characteristics of the place;
- the mobility of criminals;
- and studies of how offenders select their targets.

There are a variety of physical and social characteristics of places that make them attractive to offenders, and it is through knowledge of these and their implications for the occurrence of crimes that Environmental Criminology tries to prevent them by reducing the attractive characteristics.

The link between crime and the environment has come to influence not only criminologists, but also urban planners, architects and other researchers. Among the practical approaches involving environmental criminology, Crime Prevention Through Environmental Design (CPTED) stands out. The works of C. R. Jeffery (*Crime Through Environmental Design*, 1971) and Oscar Newman (*Defensible Space Crime Prevention Through Urban Design*, 1972) represent milestones in this field (BRANTHINGHAM'S, 1981).

An example of this type of study is the *Broken Windows* Theory, whose hypothesis is that areas with signs of vandalism and disorder are potential precursors to more serious crimes such as robberies and assaults (ANSELIN et. al., 2000).

Architect Macarena Rau (2008) assures us that crime prevention requires strategies that seek to immediately reduce crimes and the perception of fear through preventive actions, and this form of prevention must be carried out through the management of urban architecture, which can facilitate the occurrence of crimes. Vieira (2002) also highlights some variables with the potential to make the built environment more vulnerable to crime. Among the most mentioned variables are:

- lack of light
- the lack of clarity in the definition of territories;
- the lack of visibility and visual and functional connections between spaces;
- the existence of spaces that are difficult to access, preventing the victim from moving (blocked exits);

- the lack of maintenance of public areas, buildings, urban equipment and open areas (accumulation of garbage, graffiti, abandonment, etc.);
- absence of formal surveillance and pedestrian movement, which allows for natural surveillance of the spaces.

Thus, advocates of this branch (VIERA, 2002; BRANTINGHAM and BRANTINGHAM, 1981/2011; RAU, 2008; WORTLEY and MAZEROLLE, 2011, among others) claim that this range of urban design criteria would ensure that the concept of improving safety does not come at the expense of the concept of good urban quality in general.

However, some of these measures are questionable, as they can compromise the quality of life in cities and, in some cases, have the opposite effect. For example, limiting access can lead to exclusion, constant monitoring can lead to a lack of privacy and the various protective devices (bars, electrified fences, sensors, etc.) can protect your home or workplace from external threats but ignore the fact that many crimes are committed by residents and other legitimate users of the space (CLARKE, 1997). Brisman (2008) assures us: "These technologies may work in the short term, but they can also create a false sense of security and serve as a poor substitute for the proven antidote to crime: an active community, with the most human eyes on the street" (BRISMAN, 2008, p. 765).

In this sense, Klein and Walker (2005) go beyond the suggestion of transforming architectural and urban design and recommend that:

> New approaches to CPTED seek ways of socially preventing crime by raising community awareness in the urban environment, promoting street activities through fairs and other activities with residents, covering all hours of the day, in order to contribute to a collective sense of ownership and surveillance of the public sphere (KLEIN and WALKER, 2005, p. 17).

Therefore, it's not enough just to know where crime occurs and how to stop it by making individualistic changes to architecture and daily habits. We also need to think about what kind of city we are building for the future.

4 METHODS AND TECHNIQUES APPLIED TO GEOGRAPHICAL STUDIES ON CRIME AND VIOLENCE

Many studies around the world have tried to demonstrate the relationship between space and crime in the search for patterns that can reveal the mechanisms that generate the problem. However, in order to understand the relationship between space and crime, it is necessary to first understand the interaction between space and individuals, in other words, to understand the motivations behind human actions and how these actions are imprinted on space. For this reason, spaces and the phenomena that occur in them cannot be understood on their own; they need to be understood taking into account the multidimensionality of the causes and effects that surround them.

Melgaço (2005) warns that in geographic studies on crime and violence, it is necessary to consider reality in its complexity and not a fragmented reality, as has been done in analytical studies on the subject. And he assures us that in order to understand violence from a geographical perspective, it would be necessary to understand the spatial inequalities resulting from a dialectical process between crime and space.

Therefore, it is important to recognize the dialectical pairs that fit within the phenomenon: form-content and time-space. According to Santos (1985), form is constantly altered and its content also gains new functionality. In the case of the phenomenon of violence and crime, spatial forms are thought of within the logic of security, which is increasingly evident in condominiums, super-equipped buildings, in short, in the protection that imprisons. The time-space relationship is linked to the notion of the mutability of processes, since it is a spatially and temporally dynamic phenomenon.

The space-individual link can be better understood through Harvey's (1988) concept of the "geographical imagination". According to him, the geographical imagination allows the individual to recognize the role of space in their own biography, relating what they see around them and recognizing how transactions between individuals and organizations are affected by the space that separates them.

In order to define these links (space-individual-crime) more clearly, Block and Block (1995) believe it is necessary to first clarify the link between the place and the specific situation in a hierarchical way. They state that:

> The specific situation that provides the backdrop and often the mechanism for interpersonal conflict is rooted in a place, a particular small area, which reflects and affects the participants' routine activities in the short term, and plays a specific role in the conflict that takes place. Each place, in turn, is rooted in a space, a larger place that governs, in the long term, the routine activity patterns of potential participants in a conflict situation (BLOCK and BLOCK, 1995, p.146).

In this way, they suggest that the site is a point (a building, a park, an overpass) in a larger space (a census tract, a community, etc.) and that understanding its role within these hierarchical spatial levels leads to an understanding of criminal activities. This perspective leads us to spatial analysis, a research procedure widely applied in crime studies.

According to Câmara *et al.* (2004, p.2), "the emphasis of spatial analysis is to measure properties and relationships, taking into account the spatial location of the phenomenon under study in an explicit way. In other words, the central idea is to incorporate space into the analysis you wish to carry out". In addition to visualizing the spatial distribution of a problem, spatial analysis makes it possible to translate patterns established from correlations between variables. For example:

- Observing the space-time dynamics of phenomena;
- Look for correlations between the phenomenon and the socio-economic conditions of the population and between the physical characteristics of the environment such as the type of land use.

- Establish the radius of action of a crime, starting from the location of the crime, the victim's residence and the offender's residence;
- Identify whether there are preferred locations for certain types of crime, etc.

These analyses consist of a set of procedures which, in most cases, culminate in the choice of a model, usually an inferential one, i.e. one which seeks statements based on a set of representative values. Inferential models are usually presented in a one-off form or through continuous and discrete variations[6] . In the case of crime, this is a one-off phenomenon in which each criminal occurrence becomes representative and can be georeferenced.

A crucial question for this type of approach is: what is the best spatial unit for analyzing the criminal phenomenon? Paul and Patricia Brantingham (1981) already proposed a separation between levels of analysis (macro, meso and micro), emphasizing that the spatial patterns of crime differ from one level to another, i.e. different types of crime draw different territorial boundaries. In recent studies, these authors propose that local crime analysis should start from smaller units that lead to the construction of larger units and thus reflect crime patterns (BRANTINGHAM and BRANTINGHAM, 2009).

The trend towards micro-scale analysis has revealed the fundamental role of crime scene knowledge. A number of successful crime prevention efforts have been achieved using this type of approach. Modern methods of analysis have been used to identify clusters of crime around certain types of premises (bars, shopping malls, train stations, etc.) by measuring the size of the area of influence and vicinity (WEISBURD et. al., 2009).

---

[6] - For more information on this subject, see Câmara et al. (2004).

Carolyn and Richard Block (1995) clarify that space has two dimensions to consider: first, a space has attributes that are restricted to a certain level (e.g. the reputation of a neighborhood, its population, structure and level of poverty). Secondly, a space has attributes that are aggregates of characteristics of the place (for example, the number of bars or abandoned buildings per square kilometer or block of a neighborhood). In this case, we can think that the socio-spatial organization of a place can generate violence, as proposed by the Social Disorganization Theory, or the opposite, as proposed by Carrión (2008) when he states that violence generates a particular type of spatial organization, giving as an example the projection of imaginary fear that becomes a constructive element of the city . [7]

The identification of *clusters* or *outliers* leads to *Predictive Models*, which have been the great promise of this branch of activity, since it seeks to identify early warning signs over time and space, guiding crime prevention efforts through the probabilities of occurrences (GROFF and LA VIGNE, 2002).

An example of this type of technique is given by Grover et. al. (2006) when they suggest a database model, KDD (Knowledge Discovery in Database), which considers a specific area, the type of crime, the movement of the offender, repeat victimization and the analysis of *hot spots*. Another example is given by Rossmo and Rombouts (2011) who state that through *Geographic Profiling*, it is even possible to detect the location of an offender by correlating socio-spatial information.

Many studies indicate that the majority of crimes take place within a short distance of the offenders' homes. The networks (roads and paths) that connect activity nodes (home, work, etc.) also stand out in the analysis,

[7] Lindón (2007) explains that the imaginary is a social construction of the different places that make up the city. It is a process of manufacturing space that people carry out in interaction with others, guiding their spatial practices through a web of meanings called urban imaginaries.

since individuals develop standard commuting routines to their destinations, and thus contribute to defining the area that may become an offender's target (FRANK at al., 2011).

As already mentioned, crime is not distributed randomly in space; various studies have shown that there is a spatial concentration of crime (MCCORD and RATCLIFFE, 2007; WEISBURD *et al.*, 2009; BRAGA, 2010; etc.). This concentration, which had already been pointed out by Shaw and McKay (1942) in their studies on juvenile delinquency, has been reaffirmed. From this perspective, crime will be concentrated where there is a greater concentration of elements that generate criminal opportunity, or environments of differentiated criminal opportunity (SILVA, 2012).

For this reason, there is also a concentration of the phenomenon at an intra-urban level. Studies carried out in the United States show that around 50% of crimes occur in 3% or 4% of city areas (WEISBURD et. al., 2009). And this seems to be an assertion of the criminal phenomenon in other countries as well. In Brazil, the concentration of crime has already been demonstrated in various studies, characterized by the predominance of crimes against life in poor areas and crimes against property in commercial areas (FELIX, 2001; MELGAÇO, 2005; WEISEFLIZ, 2008, OLIVEIRA, 2008; SILVA, 2012; etc.).

One attempt to explain this distribution pattern may be related to locational factors such as: the circulation of people and goods in commercial areas as motivators for crimes against property, and issues related to drug trafficking in homicide cases (BEATO and REIS, 2000).

One of the processes of spatial analysis that has revealed positive associations refers to "neighborhood relations", based on the notion of "spatial dependence" given by the geographer Waldo Tobler, who says: "all things are alike, but things closer together perish more than things further apart" (*apud* CÂMARA et. al., 2004, p. 11).

Using spatial autocorrelation techniques[8] Silva (2012, p.149) found that in the city of Belo Horizonte "the violence associated with homicides, in areas where the incidence is high, tends to go beyond the local territorial limits and cause a diffusion effect to adjacent areas". In this way, it is understood that homicide rates in an area have a degree of dependence that can be related to rates in neighboring areas.

However, George Tita and Robert Greenbaum (2009) state that the correlation of events based solely on the geographical proximity of the units is insufficient to establish processes of spatial dependence, since when they studied some gangs in Pennsylvania (USA) they discovered that their influence can go beyond simple geographical contiguity through processes of social interaction.

Localized interventions raise the suspicion of crime moving to other areas. However, it is difficult both to define the extent of a possible displacement and to actually prove it, so this possibility is often put on the back burner. However, Eck and Weisburd (1995) point out that there is a growing body of evidence that this displacement is not total and is often inconsequential. They also state that recent research advocates the opposite phenomenon called "diffusion of benefits".

Some studies seek to relate crime to the type of land use. A study carried out by Anderson *et al.* (2013) in areas of the city of Los Angeles (USA) with high crime rates and similar demographic characteristics, but with different types of land use[9], revealed that mixed areas (residential and commercial) are less prone to crime than commercial-only areas.

In addition to the spatial comparison, a temporal analysis was carried out (four years) of changes in the use of some of these areas that reinforced these results. From this, the authors argue that crime can be

---

[8] - This term is derived from the statistical concept of correlation, used to measure the relationship between two random variables (CÂMARA et. al., 2004).

[9] Land use is understood here as a combination of a type of use (activity) and a type of settlement (building), and is related to *zoning* as a system of regulation.

associated with observable indicators of the built environment related to territoriality, surveillance and mobility and suggest that strategic decisions about land use could be part of an overall crime prevention strategy (ANDERSON et. al., 2013).

Studies related to violence and alcohol focus almost exclusively on the characteristics of individuals or, at most, on the characteristics of social groups in specific situations and pay little attention to place. However, a study carried out by Shermam (1992) in the city of Milwaukee (US), revealed that around 15% of bars were responsible for more than half of the crimes in bars in that city (*apud* ECK and WEISBURD, 1995).

In the city of Recife, this relationship was also positive, according to a study carried out by NEVES JR *et al.* (2013), which demonstrated the importance of bars in relation to areas that concentrate crime, leading them to the following statement:

> It is of fundamental importance to understand how the structural-physical construction of a given region contributes to explaining the phenomenon of crime, especially how aggressors choose their targets, place and time. The idea is that different configurations of the use of space in the urban area affect the way people perceive the environment around them, favoring or not the occurrence of crime. (NEVES JR *et al*, 2013, p.6)

Another example of this relationship between crime and space is given by Brisman (2008) when he refers to natural environments (forests, parks) whose vegetation can provide the offender with concealment during target selection and even when committing the criminal act. Such a relationship leads to fear of the environment itself, which has proved corrosive, because by associating the environment with crime and fear, people are unable to form bonds with natural environments and start to avoid them.

It is possible to say that this fear is generated not only in relation to natural environments, but also in urban environments with poor infrastructure, such as poor neighborhoods and slums.

4.1 CRIME MAPPING

Mapping has always been a powerful tactic for governance and surveillance. Maps are extremely important tools in crime studies, as they show not only the spatial distribution of phenomena and their correlations, but also provide guidance on what to do with the knowledge acquired, helping to devise action strategies and analyze crime patterns. The success of this type of tool in crime studies depends on a few elements that stand out in most studies using maps. These elements are: location, time, distance (HARRIES, 1999), pattern and scale (BRANTINGHAN and BRANTINGHAN, 1981-2011).

For crime analysts, location is the most important piece of information to be represented on a map, as it indicates where things have happened or where they might happen in the future. It is on the basis of this type of information that police resources are allocated.

Time becomes an important element in monitoring the evolution of crime in an area, and knowing the times when offences usually occur means organizing police shifts.

Distance becomes useful in revealing some possible types of relationships such as: the distance from the victim's home to the place where the crime took place, the maximum distance that vehicles are capable of traveling within a specific urban environment in order to respond to calls in an acceptable time and the distance that a suspect could have traveled in a specific period of time.

Patterns are seen as a powerful investigative tool, because by creating spatial models of crime occurrences and the way in which the points are configured, it is possible to see a profile of the places where crimes take place, as well as the profiles of criminals and victims. According to Brantinghan (1981), these patterns can be modeled at the neighborhood and even street level. This is where an important element of mapping comes in: scale.

The first attempts to understand the relationship between crime and the location of the crime had approaches at the "macro" level (cities, regions, states, etc.). Recently, the focus of interest has shifted to a "micro" level of approach (neighborhoods, sectors, streets, etc.) which has been tested and defended by many researchers.

In the case of crime data, the representation of areas that concentrate a high number of occurrences is called *hot spots*. The Americans Wilson and Smith (2008) draw attention to the importance of detecting crime *hot spots*, as it is through them that policing and crime prevention efforts can be directed. However, it is not simply the concentration of incidents that characterizes a *hot spot*. According to Harries (1999) not all agglomerations are considered hot zones, since the environments that help generate crime, the places where people are, also tend to constitute agglomerations. In this way, any definition of a hot zone has to be qualified, i.e. the definition of a *hot spot* depends on the regularity and predictability of occurrences in the location (ANSELIN *et al*, 2000).

The new geolocation technologies (GIS, GPS), together with wireless communication networks, have contributed to the rise of "geovigilance", which according to Crampton (2003) consists of "a mode of surveillance concerned with locations and distributions along spatial territories" (*apud* BRUNO, 2009, p.6). In recent years, this concept has led to the emergence of *online* crime maps that provide citizens with readily accessible crime patterns, whose information base comes from the population. In Brazil, the Wikicrimes and UPSEG websites (Figures 1 and 2) are examples.

Ajude sua comunidade.
Registre crimes que você teve conhecimento para que outros possam se precaver!

Carregando o mapa. Por favor, aguarde...

Login

Figure 1WikiCrimes website loading message
Source: http://www.wikicrimes.org/main.html

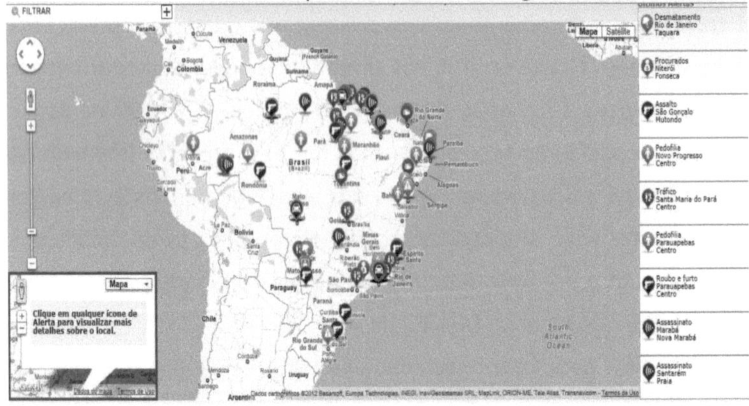

Figure 2Distribution by type of crime in Brazil on the UPSEG website
Source: http://upseg.org/mapa.upseg

Bruno (2009) asserts that these maps are a very specific case of "geovigilance" and express the state of distributed surveillance in contemporary culture, especially its participatory side.

The perspective of the advocates of this form of dissemination is that an informed public can help control crime, since the police can't be everywhere at once, and that policing will be more effective when carried out in an environment in which the public offers active support. They consider internet maps to be community resources that allow crimes committed in the city to be made visible, thus enabling city dwellers to monitor the areas they travel through and make safer travel choices.

Bruno (2009) also argues that participatory surveillance can build another regime of observation, visibility and political action carried out no longer by large centers, but by countless individuals, actions and local decisions. However, he warns that:

> Like most maps in cyberspace, crime maps are produced by ordinary individuals and not by experts in cartography, surveillance or public safety. A minority are produced by a police department, but almost all derive from the appropriation of public data by individuals or groups who are not linked to government bodies, knowledge or crime information management (BRUNO, 2009, p. 7).

This raises a worrying question: how can this information be used by the population? Although there are warnings about sanctions in case of misuse, it is necessary to consider the negative consequences for communities with a high incidence of crime or the possibility of crimes spreading to other places.

In Brazil, official data on crime and violence is monopolized by the institutions responsible for security. Although data is released in some states, there is a restriction on access to micro-data, which is justified by the fact that it does not compromise investigative processes, among other issues.

However, in some countries, more specifically in some cities, access to crime data has been opened up to the general public, as they work with the notion that disseminating information is an important prevention tool. This is the case, for example, in the city of Chicago, where the Police Department's *Chicago Crime Map* website provides information on crimes by time, location, type and even information on the criminals, as shown in the illustration below (Figure 3).

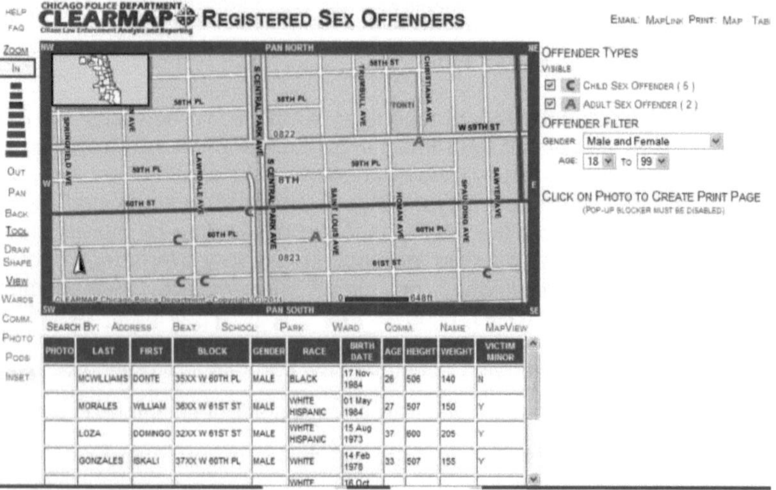

PHOTO	LAST	FIRST	BLOCK	GENDER	RACE	BIRTH DATE	AGE	HEIGHT	WEIGHT	VICTIM MINOR
	MCWILLIAMS	DONTE	35XX W 60TH PL	MALE	BLACK	17 Nov 1984	26	506	140	N
	MORALES	WILLIAM	38XX W 61ST ST	MALE	WHITE HISPANIC	01 May 1984	27	507	150	Y
	LOZA	DOMINGO	32XX W 61ST ST	MALE	WHITE HISPANIC	15 Aug 1973	37	600	205	Y
	GONZALES	ISKALI	37XX W 60TH PL	MALE	WHITE	14 Feb 1978	33	507	155	Y

Figure 3: Distribution of registered sex offenders by spatial unit of City of Chicago - 2011. Source: http://gis.chicagopolice.org/website/clearMap/viewer.htm

The maps of sexual aggression against children and adults (fig. 5), in addition to showing the distribution of the addresses of occurrences, have an associated table with data such as: name of the aggressor, gender, race, age, weight, etc.

According to information on its website, this system called "CLEARMAP" was developed to offer residents of the city of Chicago a tool to help them solve problems and fight crime and disorder in their neighborhoods, based on the system used by the department's own police officers. It is a way of involving citizens in the fight against crime by defending their territories.

However, this sharing of information with the general public can contribute to the construction of socio-spatial stigmas. Carrabine (2009) argues that:

> It remains to be seen what impact access to this information will have on "democratizing" public experiences and perceptions of crime, as well as on the authorities' abilities and desire to present crime figures in a particular way. These

> democratizing tendencies will have to be balanced alongside serious civil liberties issues. Does the resident whose home is identified as having been a crime scene have a right to privacy? (CARRABINE 2009, p. 150-151).

Therefore, it remains to be seen whether the democratization of criminal information and the sharing of public experiences and perceptions of crime will bring more benefits than harms to social life and its consequent spatial configuration.

Despite the advances and improvements in mapping techniques, we must remember that maps are built from data, which, as we have seen, should not be understood as a copy of reality. As Milton Santos said, "geometries are not geographies". The map is just an analytical tool used to represent the object of geographical study and should be used with care, to raise questions about crime and to help devise solutions to the problem, considering its limitations in representing socio-spatial phenomena.

In short, analytical techniques can help to understand many aspects of the criminal phenomenon, but their applicability has a well-established limit: they only reveal possible associations and patterns arising from a set of data and not from the reality experienced in the space. Consequently, it is necessary to go beyond the analytical and look for other methods.

4.2 THE CATEGORIES OF GEOGRAPHICAL ANALYSIS AND THE PERCEPTION OF SPACE

In geographical studies on violence and crime, some categories of analysis stand out. Territory is one of them and has been widely used in these studies because it exposes an extremely important factor in the production of space and in criminal relations: power. The classic definition of territory proposed by Claude Raffestin makes it clear how territorial delimitations are permeated by power relations when he states that:

> [...] Territory could be nothing more than the product of social actors. It is they who produce the territory, starting from the given initial reality, which is space. There is therefore a "process" of territory, when all kinds of power relations are manifested [...] (1993 p. 7-8).

Influenced by Foucault's ideas, Raffestin (1993) understands that power is present in every relational process and that there is a multiplicity of forces that make social relations dissymmetrical, i.e. where one imposes their will on the other. And it is from these power relations that territories are defined.

The relationship between the notion of territory proposed above and crime is very evident in areas dominated by drug trafficking, a delimited space where criminals operate, just as there are territories appropriated by representatives of public security, spaces considered safe due to their proximity to police stations, police stations and police posts. So it can be said that the space-crime relationship is permeated by disputes over territories where spatial domination can be exercised legitimately or illegitimately.

Milton Santos (2000) presents the concept of "used territory" which includes the result of the historical process and the material and social basis of human actions, encompassing the idea of movement and totality. According to him,

> there is no question of imposing a single definition. The content of a comprehensive geography can certainly respond to one of several theoretical lines, depending on the author's choice. But it is essential to have a coherent set of propositions, where all the elements at stake are considered in their integration and dynamism (SANTOS, 2000, p.108).

According to this conception of territory, we realize that it is not detached from its material base, i.e. its physical characteristics, even though this territory is delimited by human actions, which makes it changeable.

Marcos Aurélio Saquet (2009) explains the difference between Santos' and Raffestin's conceptions of territory. According to him, "Milton

Santos cuts space into territories without separating them, i.e. territories are in geographical space [...] Claude Raffetin does not cut space, but transforms it into a substrate for the 'creation' of territory" (SAQUET, 2009, p.77-78).

Although they are different conceptions, both emphasize the importance of the material substrate as a conditioning factor for human actions, thus corroborating the precepts of environmental criminology. Do green areas, hills, alleys, dead-end streets facilitate criminal actions? Do built-up, illuminated, monitored areas hinder such actions?

But if the construction of territory is permeated by power, then whoever holds it delimits it. This is how Marcelo Lopes de Souza (2001, p. 97) defines territory as "spatially delimited power relations that operate on a material substrate". In this way, he defends the notion that territory does not contain the material substrate itself, but only projects itself onto it. As an example of his conception, Souza cites the territories of prostitutes and transvestites in the city of Rio de Janeiro, who appropriate certain areas of the city at night, usually areas that have commercial functionality during the day. Therefore, wouldn't they be conditioned to a material substrate that provides them with discretion, which is generally overlooked in this type of activity?

When it comes to the relationship between space and crime, an approach is needed that is not solely materialistic and objective, nor one in which materiality is disregarded. Thus, the concept proposed by Santos (2000) is suitable for this type of analysis because he considers that:

> Space is the material worked on par excellence. None of the social objects has such an imposition on man, none is so present in the daily lives of individuals. The house, the workplace, the meeting points, the paths that connect these points are also passive elements that condition man's activity and command social practice (SANTOS, 1977, p. 92).

Santos considers that social actors constantly seek to adapt to the local geographical environment, while at the same time recreating strategies that guarantee their survival in places, thus characterizing a dialectical process in which space plays an active role in relation to human actions.

Another category that has proved important in the study of crime is place. With the emergence of humanist geography, subjectivity, experience and symbolism have become extremely important in understanding socio-spatial phenomena and place has emerged as the category of analysis that can make the real world intelligible (CORRÊA, 2001). According to Yi-Fu Tuan (1980), one of the exponents of this current, when space is entirely familiar to us, it becomes place. And this familiarization with place occurs through people's perceptions, attitudes and values in relation to the space they live in.

Experience and knowledge of space reveal some social mechanisms for preventing crime and violence. For example, certain socio-spatial transformations, such as risk mitigation through architecture and changes in the population's habits, can influence the occurrence of crime in a given space; on the other hand, these same changes can increase the risks in the surrounding environment by putting an end to forms of natural surveillance.

Place is the space where phenomena occur and where they can be combated. However, crime policies are directed at territories for operational reasons, since place as a category of geographical analysis can vary from a room in a house to an entire community.

Although geographical categories of analysis are promising for understanding the phenomenon under study, it must be made clear that the spatial dimension of crime is not a panacea for the ills of violent phenomena. It is a dimension that draws attention to factors that are often overlooked and which are of great relevance. Its applicability has many limitations, starting with the types of crime it relates to. For example, it is not related to corporate and white-collar crime, nor is it determinant for crimes of passion. However, it can be used both to understand the spatio-temporal distribution of criminal occurrences and to direct security policies, as well as to reveal specific conditions that make some spaces more dangerous (or more feared) than others.

Spatial analyses are generally based on data generated by police reports, which in addition to the problems already mentioned, typically only include the socio-demographic characteristics of the offender or victim and the location of the event, but do not include the physical context of the place where the crime took place (LOUKAITOU-SIDERES et. al., 2001).

Of course, neither the attributes of the place itself nor the attributes of its location in space are sufficient on their own to explain crime levels (BLOCK and BLOCK 1995), but they can reveal factors linked to the concentration of crime in certain areas and it is through this knowledge that security plans, especially their operational part, are designed.

In recent years, researchers and managers have become increasingly interested in the spatial analysis of crime. Even so, many geographic studies are limited to mapping occurrences. Some geographers, however, take a more critical perspective, trying to glimpse the socio-spatial mechanisms of the phenomenon (DINIZ *et al.*, 2003; MELGAÇO, 2005; SÁ, 2005; SOUZA 2008).

It is therefore necessary to develop alternatives for spatial evaluation that go beyond the analytical approach and mapping techniques. Because the problem emerges from a structure, but it is at the edge, in the daily manifestations of crime, that interventions are imminent. It is therefore essential to get to know the space, observe its dynamics and listen to the social actors, seeking knowledge through their experiences.

4.3 PERCEPTION OF GEOGRAPHICAL SPACE

Perception as a form of knowledge had its origins in psychology. However, as a form of perception of the environment acquired in the interrelationship between man and the world, it was first highlighted by French geography at the beginning of the 20th century, and then taken up by Anglo-Saxon geographers (CLAVAL, 1974).

Due to the materialistic and objective approach given to space throughout quantitative geographical thinking, there has been a resistance to the subjective and non-material approach that permeates the construction of this same space.

Y-fu Tuan (1980) values perception through the notion of *topophilia*, which means "the affective link between the person and the place or physical setting" (TUAN, 1980, p.5). Even when referring to the "natural physical setting", Tuan emphasizes the importance of the materiality of the place and reinforces this idea by mentioning that "the natural environment and the worldview are closely linked: the worldview, if not derived from an alien culture, is necessarily constructed from the empirical elements of the social and physical environment of a people". (op. cit., p 91). In this way, the notion of perception becomes fundamental in this work, as it is through it that we will try to understand how individuals choose between safe and unsafe spaces.

According to Tuan (1980), "perception is both the response of the senses to external stimuli and the purposeful activity in which certain phenomena are clearly registered, while others recede into the shadows or are blocked" (TUAN, 1980, p.4). Thus, trying to understand how people perceive spaces is a way of glimpsing how their mental maps of safety in places are constructed. Lindón (2007) explains:

> If a social group (whether small or as large as a nation) recognizes a place as dangerous, you are faced with a social

> construction of the place through the sense of danger assumed by that social group. This is a process in which intersubjectivity converges to attribute this meaning - which has been socially defined previously and in relation to other phenomena - to the place in question (LINDÓN, 2007, p.38).

Studies that look at issues relating to the perception of crime and violence by the population generally work with the notion of risk of victimization (RODRIGUES and FERNANDES, 2005; FELIX, 2009) without, however, spatially contextualizing factors that may be linked to the feeling of (in)security.

In an attempt to understand the immaterial mechanisms of sociospatial phenomena, human geography has tended to focus only on socioeconomic and political factors, ignoring spatial materiality and its conditioning role in human actions. According to Santos (2008), the geography of flows depends on the geography of fixes, and he explains:

At the same time as a technosphere dependent on science and technology is installed, a psychosphere is created in parallel and on the same basis. The technosphere adapts to the commands of production and exchange and, in this way, often translates distant interests, but as soon as it is installed, replacing the natural or technical environment that proceeded it, it constitutes a local fact, adhering to the place like a prosthesis. The psychosphere, the realm of ideas, beliefs, passions and the place where meaning is produced, is also part of this environment, this surrounding of life, providing rules for rationality or stimulating the imagination (SANTOS, 2008, p. 256).

In relation to violence, it is notorious for generating a "psychosphere of fear" that materializes in the form of technical objects of protection and segregation (MELGAÇO, 2007, P.214). Thus, it is understood that the material and the immaterial are inseparable and, therefore, neither of these spheres can be neglected when understanding a socio-spatial phenomenon.

Like any qualitative analysis method, the subjective approach to perception receives a lot of criticism because of the proximity between the subject and the object of study, arguing that it jeopardizes the neutrality and objectivity of scientific knowledge (MARTINS, 2004). However, Moraes explains that:

> The emphasis on subjectivity is not irreconcilable with scientific rigor. It does not exclude or replace latent meanings and unquantifiable intuitions. Content analysis, in a qualitative approach, goes beyond the manifest level, articulating the text with the psychosocial and cultural context (MORAES, 1999, p. 13).

In addition, we are interested in the "shared subjectivity" that is established as a culture and that feeds urban imaginaries[10] on the subject. According to Pyszczek (2012), the analysis of the spatial perception of insecurity caused by the fear of violence is based on three instances, as shown in the figure below:

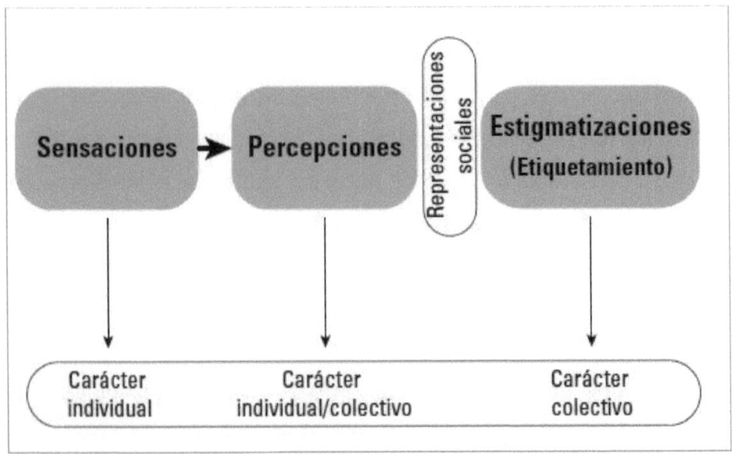

Figure 3Instances of immaterial/symbolic appropriation of geographical space. Source: Pyszczek (2012)

10 - According to Lindón (2007), these imaginaries correspond to a web of meanings that are socially constructed and reconstructed.

It thus corresponds to a process that goes from the most basic instance: sensations (which corresponds to the field of direct experiences lived by the individual) to the instance of perceptions (which includes the formation of cognitive structures that give meaning and order to spatial experience). These two instances culminate in social representations which, by perpetuating themselves in time and space, assume an "identity" inherently recognized by people and attributed to spaces and their inhabitants (urban imaginaries), and which, in the case of areas of violence, are called spatial stigmas .[11]

Lindón (2007) states that it is necessary to consider the inclusion of urban imaginaries and their role in the social construction of place as an alternative methodological strategy. He then suggests that: "the methodological work of the city scholar would begin with the production of the life accounts of different city dwellers, with a spatial component explicitly incorporated by the narrator in their discourse." (LINDÓN, 2007, p. 42). Therefore, the aim of this study is to see whether the spatial component is incorporated into people's perception of safety and how this is reflected spatially.

It is a common argument among social scientists that people's behavior is guided by their perception of the norms and values that prevail in the environment, and that these are objectified by being shared intersubjectively by a substantive group of people (VASCONCELOS, 2002). I would also add that the perception of space also plays a major role in this orientation.

[11] The word stigma, in this sense, refers to the analogy of an immaterial appropriation of space, in which certain urban sectors are associated, truly or not, with danger (PYSZCZEK, 2012).

CONCLUSION

Understanding the phenomenon of violence and crime requires a holistic view that is constantly updated, because every day we are surprised by manifestations that deviate from the norm and point to new directions and hypotheses. Furthermore, it is necessary to exercise caution when dealing with a problem that has serious implications for the lives of individuals, whose actions are shaped by the fear of violence.

Various studies have sought to uncover the relationship between space and crime. Even when the expected results are not significant, other elements emerge that highlight the importance of this relationship. It is crucial to emphasize that the study of the spatial dimension of crime is not a simple, all-encompassing geographical explanation, but rather a fundamental dimension that cannot be neglected.

Unlike most Brazilian cities, where socio-spatial inequality is evident through accentuated spatial segregation, Recife is characterized more by the process of "socio-spatial fragmentation", in which social and spatial interactions diminish or become selective and segmented by invisible borders.

This inequality is also reflected in the distribution of crime. The territorial management of security in the city can contribute to this unequal occurrence, especially in areas "privileged" by more intense monitoring. In addition, the territorial configuration of certain places, such as high rises and alleys that are difficult to access, makes police work more difficult, making them conducive to crime, although this does not mean that all these areas are criminalized.

Despite this evidence, the abandonment of these areas is noticeable compared to the territorial management of security in other parts of the city. This raises the question of how best to manage spaces that make it difficult for public authorities to act.

From the questions and assumptions initially presented, it was possible to glimpse some answers, although not definitive, which can help build an understanding of the spatial dynamics of crime and violence in these spaces. Despite the limits of the spatial approach, the contribution of this dimension to understanding a variety of crimes and to territorial security management is undeniable. It remains for us to reflect on how to promote safe spaces instead of just focusing on the safety of spaces.

REFERENCES

ADORNO, Sergio, LAMIN, Cristiane. Fear, Violence and Insecurity. In: LIMA, Renato S. de; PAULA, Liana de (Orgs.). **Public security and violence**: is the state fulfilling its role? São Paulo: Ed. Contexto, 2006.

ADORNO, Sérgio. Socio-economic exclusion and urban violence. **Sociologias**, Porto Alegre, ano 4, n 8, p. 84-135, jul/dez. 2002.

ANDERSON, James M. *et al.* Reducing crime by shaping the built environment with zoning: an empirical study of Los Angeles. **University of Pennsylvania Law Review**, v.161:699, 2013.

ANSELIN, Luc et. al. Spatial Analyses of Crime. **Criminal Justice**, v. 4, 2000, pp. 213-262.

ARENDT, Hannah. **On Violence**. Rio de Janeiro: Vozes. 1994.

BARBOSA, Andrea M. F. *et al.* Homicide and living conditions: the situation in the city of Recife, PE. **Epidemol Serv. Saúde**, Brasília 20(2), p.141-150, abr.-jun., 2011.

BAUMAN, Zygmunt. **Liquid Fear**. Rio de Janeiro: Jorge Zahar, 2008.

BEATO, Cláudio C. **Crime and Cities**. Belo Horizonte: UFMG, 2012.

BEATO, Cláudio C; REIS, Ilca A. Inequality, Socio-Economic Development and Crime. In: HENRIQUES, R. (Org.). **Inequality and Poverty in Brazil**. Rio de Janeiro: IPEA, 2000.

BITOUN, Jan. What the human development indices reveal. In: **Human Development in Recife**: municipal atlas. Recife, 2005, CD-ROM.

BLOCK, Richard. L.; BLOCK Carolyn R. Space, Place and Crime: Hot Spot Areas and Hot Places of Liquor-Related Crime. In: ECK, John E; WEISBURG, David (Eds). **Crime and Place**: Crime Prevention Studies, v. 4. Criminal Justice Press, Monsey, New York, p. 145 - 146, 1995.

BRAGA, Anthony A. WEISBURD, David L. Empirical Evidence on the Relevance of Place in Criminology. **Journal of Quantitative Criminology**, v. 26, 1 ed. p. 1-6, 2010. Available at: <http://link.springer.com/article/10.1007%2Fs10940-009-9088-4?LI=true> Accessed on: October 7, 2011.

BRANTINGHAN, Paul. J. BRANTINGHAN, Patricia. L. Environmental Criminology. In: JACOBY, Joseph. **Classics of Criminology**, Long Grove, Waveland, Chap. 10, p. 61-70, 1981- 2004.

BRANTINGHAN, Paul. J. Crime Pattern Theory. In: MAZEROLLE, Lorraine (Eds.). **Environmental Criminilogy and Crime Analysis**. New York: Routledge, 2011.

BRAZIL, Decree-Law No. 3914 of December 9, 1941. Available at: <http://www.planalto.gov.br/ccivil_03/decreto-lei/Del3914.htm> Accessed on: 11/08/2012.

BRISMAN, Avi. Crime-Environment Relationships and Environmental Justice. **Seattle journal for social justice**. v. 6, 2 ed., p. 727- 817, 2008.

BRUNET, Júlio F. G. *et al.* Predictive factors of violence in the Metropolitan Region of Porto Alegre. **Revista Brasileira de Segurança Pública**, ano 2, ed. 3, jul.-ago. 2008.

BRUNO, Fernanda. Crime Maps: distributed surveillance and participation in cyberculture. **Revista da Associação Nacional dos Programas de Pós-Graduação em Comunicação - E-compós**, Brasilia, v. 12, n°2, p. 1-16, mai-ago. 2009.

CÂMARA, Gilberto *et al.* Spatial Analysis and Geoprocessing. In: DRUCK, S; CRVALHO, M. S; CÂMRA, G; MONTEIRO, A. V. M. (eds). **Spatial analysis of geographic data**. Planaltina, DF: Embrapa Cerrados, 2004.

CAMPOS COELHO, Edmundo C. **A Oficina do Diabo e Outros Estudos sobre Criminalidade**. Rio de Janeiro: Record, 2005.

CAMPOS COELHO, Edmundo C. Sobre sociólogos, pobreza e crime. **Dados**, Rio de Janeiro, v. 23, n. 3, p.377-383, 1980.

CARRABINE, Eamonn. Crime, Place and Space. In: CARRABINE, Eamonn et. al. (Eds.) **Criminology**: a sociological introduction. 2nd edition, New York, NY: Routledgy, p. 137-153, 2009.

CARREIRA, Denise. **National Rapporteur for the Human Right to Education**: Education in Brazilian Prisons / Denise Carreira and Suelaine Carneiro, São Paulo: Plataforma DhESCA Brasil, 2009.

CARRIÓN, Fernando. Urban violence: a city issue. **Revista Latinoamericana de Estudios Urbanos e Regionales - EURE**, Chile, v. 34, n. 103, p. 5-26, dec. 2008.

CLARKE, Ronald V. **The Theory of Crime Prevention Through Environmental Design**, 1997. Available at: < www.e-doca.net > Accessed on: July 4, 2011.

CLARKE, Ronald V.; CORNISH, Derek B. Modeling Offenders' Decisions: A Framework for Research and Policy. In: JACOBY, Joseph. **Classics of Criminology**, Long Grove, Waveland, Chap. 15, p. 109 - 118, 1985 - 2004.

CLAVAL, Paul. Géographie et la percepción de l'espace. In: **Espace Géographique**, Tome 3, nº 3, p. 179-187, 1974.

COHEN, Lawrence E.; FELSON, Marcus. Social Change and Crime: A Routine Activity Approach. In: JACOBY, Joseph. **Classics of Criminology**, Long Grove, Waveland, Chap. 9, p. 52-60, 1979 - 2004.

COHN, Ellen G. Weather and Crime. **British Journal of Criminology**, v. 30, n°1, p. 51-64, Winter, 1990. Available at: < http://ibis.geog.ubc.ca/courses/geob370/students/class07/crime_weather/misc/weather_and_crime.pdf> Accessed on: 07 Oct. 2011.

CORRÊA, Roberto L. Space: a key concept in Geography. In: CASTRO, I. E.

CORRÊA, Roberto L. Formas simbólicas e espaço: algumas considerações. GEOgrafia - Year IX - No 17 - 2007

COSTA, Inês E.R; LUDERMIR, Ana B; SILVA Isabel A. Differentials in mortality due to violence against adolescents according to Stratum of Condition of Life and Race-Color in the city of Recife. **Revista Ciência & Saúde Coletiva**, 14(5), p.1781-1788, 2006.

CRUZ, Luciana. SÁ, Alcindo J. **Medo Urbano e suas novas formas geográficas**. Recife, University, 2011.

DENZIN, Norman K. **The research act**: a theoretical introduction to sociological methods. New York: McGraw-Hill, 1978.

DINIZ, Alexandre M. A.; BATELLA, Wagner B. Spatial approaches to the study of violent crime. In: II Simpósio Internacional sobre Cidades Médias - Dinâmica Econômica e Produção do Espaço, 2006, Uberlândia. **Anais...**Uberlândia, v. único, p.1-20, 2006.

DINIZ, Alexandre M. A.; NAHAS, Maria I. P.; MOSCOVITCH, Samy K. Spatial analysis of urban violence in Belo Horizonte: a methodological proposal based on georeferenced information and indicators. **Caderno de Geografia.** V. 13, n. 20, p.39-56, 2003.

DUARTE, Teresa. **The possibility of research 3**: reflections on (methodological) triangulation. Center for Research and Studies in Sociology - CIES e-Working Paper, nº 60, 2009. Available at: < http://www.cies.iscte.pt/destaques/documents/CIES-WP60_Duarte_003.pdf> Accessed on: June 17, 2013.

DURAND, Gilbert. **The Anthropological Structures of the Imaginary: an introduction** to general archetypology. Translated by Hélder Godinho. São Paulo: Martins Fontes, 1997.

ECK, John E; WEISBURD, David. Crime Places in Crime Theory. In: ECK, John E; WEISBURD, David. **Crime and Place**: Crime Prevention Studies, v. 4, Criminal Justice Press, Monsey, New York, p. 1-33, 1995.

FÁVERO, L. L. Belfiore, P. P. Chan, B. L. Silva, F. L. (2009). **Data Analysis**: Multivariate modeling for decision making. Rio de Janeiro: Elsevier, 2009.

FELIX, Sueli A. A Geografia das Ofensas: análises dos espaços de crimes, criminosos e das condições de vida da população de Marília - SP. **Scientific Report** - UNESP, 2001. Available at: < http://www.levs.marilia.unesp.br/GUTO/relatorios/relat_fase1.pdf> Accessed on: Nov. 20, 2011.

FELIX, Sueli A. Crime, Fear and Perceptions of Insecurity. Perspectivas, São Paulo, v. 36, p. 155-173, jul./dez. 2009.

FERREIRA, Norma S. de A. Research called "State of the Art". **Educação & Sociedade**, ano 23, nº 79, p. 257-272, aug/2002.

FIGUEIREDO FILHO *et al.* Principal component analysis for the construction of social indicators. **Revista Brasileira de Biometria**, v.31, n.1, 2013, p.61-78. Available at:

http://jaguar.fcav.unesp.br/RME/fasciculos/v31/v31_n1/A5_Dalson_Ranulfo.pdf Accessed on: Feb. 24, 2014.

FRANCO, Maria S. de C. O código do Sertão. In: **Homens livres na sociedade escravocrata**. 4th ed. São Paulo: UNESP, 1997.

FRANK, R. *et al.* Power of Criminal Attractors: modeling the pull of activity nodes. **Journal of Artificial Societies and Social Simulation - JASSS**, v.14. 27 pp., Jan 2011, Available at: < http://www.sfu.ca/~vdabbagh/jasss.pdf> Accessed on : July 04, 2011.

GOMES, P. C. da C. CORREA, R. L.(orgs). **Geography**: Concepts and Themes, Rio de Janeiro: Bertrand Brasil, 2001.

GROFF, Elizabeth R; LA VIGNE, Nancy G. Forecasting the future of predictive crime mapping. **Crime Prevention Studies**, v. 13, pp.29-57, 2002.

GROVER, Vika *et al.* Review of Current Crime Prediction Techniques. University of Portsmouth, School of Computing, **Research Report**, 2006. Available at: < www.maxbramer.org.uk/papers/crime_prediction.pdf> Accessed on: July 4, 2011.

GURGEL, Wildoberto B. Triangulation in debate: considerations on the Minayan model of evaluation by triangulation of methods. **Ciências Humanas em Revista**, São Luiz, v. 5, nº 1, jul. 2007

HARRIES, Keith. **Mapping crime**: principles and practice. Washington DC: National Institute of Justice, U.S. Department of Justice; 1999.

HARVEY, David. **Social Justice and the city**. Oxford: Basil Blackwell, 1988.

IBGE - Brazilian Institute of Geography and Statistics. **Demographic Census**: results of the universe. Rio de Janeiro, 2010. Available at: <http://censo2010.ibge.gov.br/resultados Accessed on: 18 Dec. 2013.

IBGE - Brazilian Institute of Geography and Statistics. **Victimization Survey**, PNAD, 2009. Available at: < http://biblioteca.ibge.gov.br/visualizacao/monografias/GEBIS%20-%20RJ/pnadvitimizacao.pdf> Accessed on October 18, 2011.

KLEIN, Jayne, WALKER, Ryan C. Statutory and non-statutory approaches to crime prevention through environmental design, **Planning Quarterly**, 159, p. 15-17, (December) 2005.

LAVILLE, Christian; DIONNE, Jean. **The construction of knowledge**: a manual of research methodology in the human sciences. Belo Horizonte: Editora UFMG, 1999.

LIMA, Renato S. Redeeming the value of quality information. **Anuário Brasileiro de Segurança Pública**, Fórum Brasileiro de Segurança Pública, year 5, p. 8-11, 2011a.

LIMA, Renato S; RATTON, José L. (Orgs) **As ciências sociais e os pioneiros nos estudos sobre crime, violência e direitos humanos no Brasil**. São Paulo: ANPOCS, 2011b.

LINDÓN, Alicia. Urban imaginaries and geographical constructivism: spatial holograms. **Revista Eure**, v. 33, n°9, p.31-46, Santiago de Chile, Aug. 2009.

LOUKAITOU-SIDERES, A. *et al.* Measuring the effects of built envinment on bus stop crime. **Emvironment and Planning B**, 28 (2), p. 255-280, 2001. Available at: < http://www.uctc.net/papers/419.pdf> Accessed on: 07 Oct. 2011.

LUNA, Fabián G. Spatialization of violence in Latin American cities: a theoretical approach. **Revista Colombiana de Geografía**, v. 22, nº 1, p. 169-186, jan.-jun., 2013.

MACEDO, Adriana C; PAIM, Jair N. S; SILVA, Lígia M. V; COSTA, Maria C. N; Violence and social inequality: homicide mortality and living conditions in Salvador, Brazil. **Rev. Saúde Pública** [online], vol.35, n.6, p. 515-522, 2001.

MACHADO DA SILVA, Luiz A. **Vida sob Cerco**: violência e rotina nas favelas do Rio de Janeiro. São Paulo: Nova Fronteira, 2008.

MARTINS, Heloisa H. T. S. Metodologia qualitativa de pesquisa. **Educação e Pesquisa**, São Paulo, v. 30, nº 2, p. 289-300, mai/ago., 2004.

MCCORD, Eric S.; RATCLIFFE, J. H. Intensity value analysis and the criminogenic effects of land use features on local crime patterns. **Crime Patterns and Analysis**, 2(1), p. 17-30, 2009.

MELGAÇO, Lucas M. **A Geografia do frrito**: dialética espacial e violência em Campinas - SP. Dissertation (Master's in Geography) University of São Paulo, SP, 2005.

MELGAÇO, Lucas M. **Da Psicosfera do Medo à Tecnosfera da Segurança**. In: SÁ, Alcindo J. (org.) "Por uma geografia sem cárceres públicos ou privados", Recife: [s.n.], 2007.

MELO. Patrícia B. **Stories the media tells**: the discourse of violent crime and the cultural trauma of fear. Recife: Ed. Universitária/UFPE, 2010.

MERTON, Robert K. Social Structure and Anomie. In: JACOBY, Joseph. **Classics of Criminology**, Long Grove, Waveland, Chap. 27, p. 214-223, 1938 - 2004.

MINAYO, Maria C. de S. The challenge of social research. In: Minayo, M. C. S; DESLANDES, S. F; GOMES, R. (orgs). **Pesquisa Social**: teoria, método e criatividade, 28 ed., Petrópolis, RJ: Vozes, 2009.

MISSE, Michel. **Crime and Violence in Contemporary Brazil**: studies in the sociology of crime and urban violence, Rio de Janeiro: Lumen Juris, 2006.

MORAES, Roque. Content analysis. **Revista Educação**, Porto Alegre, v. 22, n. 37, p. 7-32, 1999. Available at: < http://cliente.argo.com.br/~mgos/analise_de_conteudo_ moraes.html> Accessed on: 21 Jan. 2014.

NEVES JR, Edivaldo C.; PAEZ, Antonio; MENEZES, Tatiana A. Detecting spatial clusters of violent crime: the importance of bars and churches. **II Encontro Pernambucano de Economia**, Nov. 2013.

NOSSA, Paulo N. S. Data analysis methods and techniques frequently applied in Health Geography. In: NOSSA, Paulo N. S. **Abordagem geográfica da Oferta e Consumo de Cuidados de Saúde**, 2005, 329 p. Thesis (Doctorate in Geography) Institute of Social Sciences - University of Minho, 2005.

OLIVEIRA, Giovane R. Senso Comum, Pobreza e Criminalidade**, International Colloquium (Des) envolvimentos contra a pobreza: medções teóricas, técnicas e políticas.** Montes Claro, August 21-23, 2008. Available at: <http://www.coloquiointernacional.unimontes.br/2008/arquivos/118giova nerodrigu esdeoliveira.pdf> Accessed on: October 7, 2011.

OSGOOD, Wayne D.; CHABERS, Jeff M. Social Disorganization Outside The Metropolis: An Analysis of Rural Youth Violence. ***Criminology,*** 38:81-115, 2000.

PAIXÃO, Antônio L. Crime, Social Control and the Consolidation of Democracy: the metaphors of citizenship. In: REIS, Fabio W; O'DONNEL, Guillermo (eds). **Democracy in Brazil**: dilemmas and perspectives, Rio de Janeiro: Vértice, 1988.

PEDRAZZINE, Yves. **The Violence of Cities.** Translated by Giselle Unti, Petrópolis, RJ: Vozes, 2006.

PERALVA, Angelina. **Violence and Democracy**: The Brazilian Paradox. São Paulo: Paz e Terra, 2000.

PERLMAN, J. E. **O Mito da marginalidade:** favelas e políticas no Rio de Janeiro. Rio de Janeiro: Paz e Terra, 1977.

PYSZCZEK, Oscar L. Los espacios subjetivos del miedo: construcción de la estigmatización espacial en relación con la inseguridad delictiva urbana. **Revista Colombiana de Geografía**, v. 21, n°1, jan-jun. 2012.

RAFFESTIN, Claude. **For a Geography of Power**. São Paulo: Ed. Ática, 1993.

RATTON JR, José L. A. Rationality, Politics and the Normality of Crime in Émile Durkheim. **Argumentum**, Recife, v.1, p. 111-130, 2005. Available at: < http://www.maristaspe.com/argumentum/volume1/ratton.pdf>. Accessed on: Dec. 15, 2009.

RAU, Macarena. Preventing crime through environmental design. In: **Seminar on Safe Urban Spaces**, 2008, Recife. Available at: < http://www.chs.ubc.ca/consortia/events/eventsP-20080916.html> Accessed on: 10 Nov. 2008.

RODRIGUES, Corine D.; FERNANDES, Rodrigo A. Fear of Crime: Perception or reality? a comparative analysis of perceived risk and objective risk of local and non-local victimization. In: XII Congresso Brasileiro de Sociologia, Belo Horizonte, MG, May 31 to June 3, 2005.

ROH, Sunghoon; CHOO, Tae M. Looking Inside Zone V: Testing Social

Disorganization Theory In Suburban Areas. ***Western Criminology Review,*** 9(1), 1-16, 2008.

ROSSMO, Kim; ROMBOUTS, Sacha. Geographic profiling. In: WORTLEY, Richard; MAZEROLLE, Lorraine (Eds.). **Environmental Criminology and Crime Analysis**. New York: Routledge, 2011.

SA, Alcindo José. Solid" fear: the spread of security apparatuses on the outskirts of Recife and their impact on intra-urban morphologies. In: SÁ, Alcindo J. (org). **Dos espaços do medo a psicoesfera da civilidade, a premência de uma nova economia política/territorial**. Ed, Universitária/UFPE: Recife, 2010.

SA, Alcindo José. **Brazil Incarcerated**: from prisons outside prisons to prisons inside prisons: a geography of fear. Recife: Universitária-UFPE, 2005.

SANTOS, Milton. Sociedade e Espaço: a formação social como teoria e como método. **Boletim Paulista de Geografia**, São Paulo, nº 54, p. 81-99, jun. 1977.

SANTOS, Milton. **Space and method**. São Paulo: Nobel, 1985.

SANTOS, Milton. **The active role of geography**: a manifesto. Florianópolis: Estudos Territoriais Brasileiros, LABOPLAN, July 2000.

SANTOS, Milton. **The Nature of Space:** Technique and Time. Reason and Emotion. São Paulo: Edusp, 2008.

SAPORE, Luís F. **Segurança Pública no Brasil**: desafios e perspectivas. Rio de Janeiro: FGV, 2007.

SAQUET, Marcos A. Towards a territorial approach. In: **Territories and Territorialities**: theories, processes and conflicts. In: SAQUET, Marcos A; SPOSITO, Eliseu S. (eds). São Paulo: Expressão Popular - UNESP, 2009.

SAURET, Gerard. Homicide counting policies in Brazil: the cases of São Paulo, Rio de Janeiro, Minas Gerais and Pernambuco. In: SAURET, Gerard V (org.) **Statistics for life**: the collection and analysis of criminal information as an instrument of lethal violence. Recife: Bagaço, 2012.

SHAW, Clifford R.; MCKAY Henry D. Juvenile Delinquency and Urban Areas. In: JACOBY, Joseph. **Classics of Criminology**, Long Grove, Waveland, Chap. 4, p. 19-25, 1942 - 2004.

SICHE, R. *et al.* (2007). Indices versus Indicators: conceptual precision in the discussion of country sustainability. **Ambiente & Sociedade**, v. 10, n. 2, p. 137-148, jul.-dez. Available at: <http://www.scielo.br/scielo.php?script=sci_arttext&pid=S1414-753X2007000200009> Accessed on May 26, 2014.

SILVA, Braulio F. A. **Disorganization, opportunity and crime**: an "ecological" analysis of homicides in Belo Horizonte. Belo Horizonte, MG. Thesis (Doctorate in Sociology) Federal University of Minas Gerais, MG, 2012.

SOUZA, Marcelo L. **Fobópole**: O Medo Generalizado e a Militarização da Questão Urbana, Rio de Janeiro: Bertrand Brasil, 2008.

SOUZA, Marcelo L; O território: sobre espaço e poder, autonomia e desenvolvimento, in: CASTRO, Iná E. GOMES, Paulo C. da C. CORREA, Roberto L.(orgs) **Geografia**: Conceitos e Temas, Rio de Janeiro, Bertrand Brasil, 2001.

SUTHERLAND, Edwin H. White-Collar Criminality. In: JACOBY, Joseph. **Classics of Criminology**, Long Grove, Waveland, Chap. 3, p. 13-18, 1940 - 2004.

TITA, George E; GREENBAUM, Robert T. Crime, Neighborhoods, and Units of Analysis: putting space in its place. In: WEISBURD et. al. (eds.). **Putting Crime in its Place**: Units of Analysis in Geographic Criminology. Springer, New York, p. 145-170, 2009.

TUAN, Yi-fu. **Topophilia**: a study of the perception, attitudes and values of the environment. São Paulo: Difel, 1980.

VASCONCELOS, Eduardo M. **Complexidade e pesquisa interdisciplinar**: Epistemologia e metodologia operativa. Petrópolis: Vozes, 2002.

VIEIRA, Liése B. **The Influence of Built Space on the Occurrence of Crime in Housing Estates.** 2002. 310f. Dissertation (Master's in Urban and Regional Planning) - Faculty of Architecture and Urbanism, Postgraduate Program in Urban and Regional Planning, UFRGS, Porto Alegre, 2002.

WAISELFISZ, Julio J. **Mapa da Violência 2014**: os jovens do Brasil. FLASCO Brasil: Brasília, 2014. Available at: <www.juventude.gov.br/juventudeviva> Accessed on: 04 Aug. 2014.

WEISBURD et. al. (eds.). **Putting Crime in its Place**: Units of Analysis in Geographic Criminology. Springer, New York, 2009.

WILSON, Ron; SMITH, Kurt. **What is applied Geography for the study of crime and public safety?** A Quarterly Bulletin of Applied Geography for the Study of crime & Public Safety, Whashington, D.C., Office of Community Oriented Policing Services, v.1, Ed. 1, Mar. 2008. Available at: <http://www,cops.udoj.gov/files/ric/Publications/GPS _Newsletter_March_08.txt>. Accessed on: Jan. 14, 2010.

WONG, Carlin. **Clifford R. Shaw and Henry D. McKay**: The Social Disorganization Theory. Center for Spatially Integrated Social Science: Back to Classics. Santa Barbara: Regents of University of California, 2001. Available at: < http://www.csiss.org/classics/content/66> Accessed on: 09 Apr. 2011.

WORTLEY, Richard; MAZAROLLE, Lorraine. **Envronmental Criminology and Crime** Analysis. London and New York: Routledge, 2011.

ZALUAR, Alba. Unfinished democratization: failure of public security. **Revista Scielo**, Estudos Avançados, 21 (61), 2007. Available at:< http://www.scielo.br/scielo.php?pid= S0103-40142007000300003&script=sci_arttext> Accessed on: 19 Dec. 2011.

ZALUAR, Alba. Violence and crime. In: Miceli, S. (ed.). **What to read in the social sciences in Brazil**. São Paulo: Sumaré; ANPOCS, *p*. 13-107, 1999.

Printed by Books on Demand GmbH, Norderstedt / Germany